RESOURCE ALLOCATION IN MULTIUSER MULTICARRIER WIRELESS SYSTEMS

T0142942

RESOURCE ALLOCATION IN MULTIUSER MULTICARRIER WIRELESS SYSTEMS

Ian Wong
The University of Texas at Austin
Department of Electrical and Computer Engineering
Austin, Texas

Brian Evans
The University of Texas at Austin
Department of Electrical and Computer Engineering
Austin, Texas

 Springer

Ian Wong
The University of Texas at Austin
Department of ECE
Austin, TX 78712
ian.wong@ieee.org

Brian Evans
The University of Texas at Austin
Department of ECE
Austin, TX 78712
bevans@ece.utexas.edu

Resource Allocation in Multiuser Multicarrier Wireless Systems
by Ian Wong and Brian Evans

ISBN-13: 978-1-4419-4522-8   e-ISBN-13: 978-0-387-74945-7

Printed on acid-free paper.

9 8 7 6 5 4 3 2 1

springer.com

For our lovely wives, Lynn and Mouna

Preface

Next-generation broadband wireless standards, e.g. IEEE 802.16e and Third Generation Partnership Project – Long Term Evolution (3GPP-LTE), use Orthogonal Frequency Division Multiple Access (OFDMA) as the preferred physical layer multiple access scheme, esp. for the downlink. Due to the limited resources available at the base station, e.g. bandwidth and power, intelligent allocation of these resources to the users is crucial for delivering the best possible quality of service (QoS) to the consumer with the least cost.

The problem of allocating time slots, subcarriers, rates, and power to the different users in an OFDMA system has been an area of active research in recent years. Previous research efforts in OFDMA resource allocation have typically focused on maximizing instantaneous performance, i.e. the allocation decisions are performed for the current time instant subject to the current resource constraints, which is unable to fully utilize the time-varying nature of the wireless channel to improve the communication performance of the system. This book focuses instead on maximizing time-averaged rates, allowing us to exploit the temporal dimension to improve performance.

Furthermore, due to the difficult combinatorial nature of the problem, many researchers in the past have focused on developing sub-optimal heuristic algorithms. This book proposes a unified algorithmic framework based on dual optimization techniques that have complexities that are linear in the number of subcarriers and users, and that achieve negligible optimality gaps in standards-based numerical simulations. Adaptive algorithms based on stochastic approximation techniques are also proposed, which are shown to achieve similar performance with even much lower complexity.

Finally, it was assumed in previous work that perfect channel state information (CSI) is available at the transmitter, which is quite unrealistic due to inevitable channel estimation errors and feedback delay. This book develops algorithms assuming that only imperfect CSI is available, such that allocation decisions are made while explicitly considering the error statistics of the CSI.

Austin, TX
June 2007

Ian Wong
Brian Evans

Contents

List of Tables

List of Figures

1

Introduction

Nikola Tesla wrote about his wireless system in 1900 [4]

"I have no doubt that it will prove very efficient in enlightening the masses, particularly in still uncivilized countries and less accessible regions, and that it will add materially to general safety, comfort and convenience, and maintenance of peaceful relations."

Fast forward to 2007, we can only marvel at the accuracy of his prediction. Today, developing countries in the Asia-Pacific are the fastest growing adopters of cellular wireless technology. According to market research from Frost and Sullivan, the Asia-Pacific cellular subscriber base reached 819.5 million at the end of 2006, and is forecast to reach 1.68 billion by the end of 2012, resulting in a compounded annual growth rate of 10.8% [5]. Furthermore, the most recent research in wireless communications are trending towards systems that benefit public-safety, environmental health, and military applications [6]. It is without a doubt that wireless communications has indeed contributed significantly to the safety, comfort, and convenience in almost every aspect of modern society. In this book, we focus on one of the most significant wireless technologies that impact our lives today: wireless voice and data communications.

We begin this introduction chapter by briefly outlining the evolution of wireless voice and data communication systems in Sec. 1.1. We discuss various multiple access schemes that have been proposed in the past, and observe that the multiple access scheme that achieves high data rates while being resilient to the harsh wireless fading environment is orthogonal frequency division multiple access (OFDMA). This scheme is thus the preferred method for sharing the wireless spectrum for next-generation wireless systems, and is introduced in Sec. 1.2. Due to the ever-increasing demand for reliable voice and data communications, the efficient allocation of resources to the users in an OFDMA system is a crucial problem to solve. This is a very difficult problem whose efficient solution has eluded researchers in the past, and is the primary focus of this book. The summary of my book is provided in Sec. 1.3, where we present the thesis statement, summarize the contributions, and discuss the overall organization. We then end this chapter in Sec. 1.4 with a listing of the most commonly used acronyms in this book for easy reference.

1.1 Next-generation Wireless Communication Systems

As of Dec. 2006, the worldwide cellular telephony subscriber base has reached 2.69 billion customers [7], roughly 41% of the world population of 6.53 billion. This is projected to increase to 4.3 billion by 2011, roughly 62% of the projected world population at that time of 6.92 billion [8]. On the other hand, the popularity of the Internet is also growing tremendously, with 1.1 billion worldwide users as of March 2007, approximately a 200% growth since 2000 [9]. Given these trends in our ever more connected world, it is conceivable why wireless communications is moving from providing simple voice service to delivering heterogenous business and consumer data-centric applications. In fact, global revenues from mobile data services exceeded $100 billion in 2005, and is projected to reach $166 billion by 2010 based on current growth trends [1].

It is not surprising then that current and future mobile devices are also evolving into highly integrated multi-functional business and entertainment gadgets, combining wireless Internet and email portal, electronic organizer, still and video camera, MP3 audio player, portable gaming console, etc., into a single device. Subscribers are expecting access to information, communication, and entertainment anytime and anywhere. The services envisioned for this growing market require increased data rates, wider coverage, and improved link reliability of the wireless network (see Table 1.1 for example applications and their typical data rate requirements). Thus, efficient use of the scarce resources, e.g. spectrum, power, and time, are of paramount importance.

Table 1.1. Wireless data applications and their required data rates [1]

Application	Data rate
Microbrowsing (Wireless Access Protocol (WAP))	8 − 32 kbps
Multimedia Messaging	8 − 64 kbps
Audio and Video Streaming	32 − 384 kbps
Video Telephony	64 − 384 kbps
General Purpose Web Browsing	32 kbps- > 1 Mbps
Enterprise Applications (e.g. database access)	32 kbps- > 1 Mbps

Next-generation wireless standards, e.g. Third-generation partnership project-long term evolution (3GPP-LTE) and IEEE 802.16e, are poised to meet this future demand for wireless services. Using state-of-the-art wireless communications technologies, these two standards are expected to deliver peak data rates of up to 100 Megabits per second (Mbps) to users traveling at vehicular

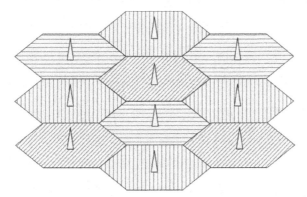

Fig. 1.1. Cellular wireless communication system with hexagonal cells. Different shading patterns of the cells indicate different sets of frequency allocations.

speeds. These two emerging wireless communication standards evolved from different technological camps: 3GPP-LTE from the voice-centric cellular network architecture, and IEEE 802.16e from the data-centric broadband access network architecture (e.g. digital subscriber lines and cable Internet). Interestingly, both actually share some striking similarities in their technological choices, particularly in their choice of physical layer transmission schemes. The following two subsections shall discuss these two emerging standards, where we briefly outline their technological evolution.

1.1.1 Evolution of Cellular Standards to 3GPP-LTE

In the late 1960s, Bell laboratories first developed the concept of cellular wireless communications [10], wherein spectrum within a geographical region can be reused by breaking the region into a tessellation of hexagonal "cells." Each cell is assigned a set of frequencies, and, due to the physical phenomenon of radio strength attenuation with increasing distance, these frequencies can be reused either by the adjacent cells, or in the second tier of cells, and so on. Fig. 1.1 shows an example cellular wireless communication system where adjacent cells do not occupy the same set of frequency channels. This cellular concept, coupled by the developments in reliable solid state radio frequency (RF) hardware, ushered in the modern wireless communications era.

First Generation (1G)

In the 1980s, the first generation of cellular networks (1G) were deployed in Japan, the United States, and Europe [11]. These 1G networks used analog frequency modulation (FM), where each subscriber making a call was assigned a separate downlink and uplink FM channel. This method of spectrum sharing, wherein the users and the transmission direction are assigned disjoint

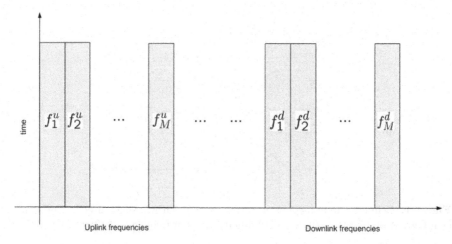

Fig. 1.2. Frequency division multiple access (FDMA) with frequency division duplexing (FDD).

partitions in frequency, is called frequency division multiple access (FDMA) with frequency-division duplexing (FDD). In FDMA and FDD, it is important to separate the channels sufficiently such that inter-channel interference can be mitigated using practical filters. Fig. 1.2 shows a typical FDMA with FDD setup, where users are assigned an uplink and downlink frequency channel, e.g. f_m^u and f_m^d for the entire duration of the connection.

Second Generation (2G)

As the number of subscribers grew, it was clear that analog technology could not use the spectrum efficiently enough to sustain the growth in popularity of cellular telephone service. Thus, in the early 1990s, second generation (2G) cellular networks that use digital modulation were developed. The most widely used 2G standard in the world today, with approximately 2 billion subscribers, is the Global System for Mobile Communications (GSM). GSM similarly uses separate sets of uplink and downlink frequencies (FDD), but users share the spectrum using separate time slots. This method of spectrum sharing is called time-division multiple access (TDMA), and is shown in Fig. 1.3 with FDD. GSM was also designed to support low rate data services of up to 9.6 kbps.

Third Generation (3G)

In the late 1990s, fueled by the surge in popularity of the Internet, consumer demand for wireless data services has likewise increased. Thus, third generation (3G) cellular standards were designed to support the following minimum data rates in the various mobility environments:

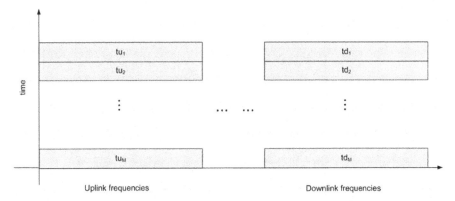

Fig. 1.3. Time division multiple access (TDMA) with frequency division duplexing (FDD).

1. Vehicular: 144 kbps
2. Pedestrian: 384 kbps
3. Indoor Office: 2 Mbps
4. Satellite: 9.6 kbps

The most popular 3G standards, i.e. Universal Mobile Telecommunications System (UMTS) which evolved from GSM, and cdma2000 which evolved from the IS-95 2G standard, are both based on FDD with code division multiple access (CDMA) technology. CDMA allows different users to transmit at the same time and frequency, but using different "codes." Fig. 1.4 shows a diagram of CDMA where there are separate uplink and downlink frequencies (FDD), and users are separately assigned different codes, but use the same time and frequency blocks. These codes, when designed to be orthogonal, i.e. "non-interfering", can then effectively separate the users from each other, allowing simultaneous links to be maintained with minimal interference [10].

Fourth generation and beyond (3GPP-LTE)

3GPP-LTE is a new wireless standard currently under development by the 3GPP (http://www.3gpp.org), with a planned initial deployment in 2009. LTE is envisioned as the fourth generation cellular standard, and is aligned with existing third-generation deployments, e.g. UMTS. 3GPP-LTE uses orthogonal frequency division multiple access (OFDMA) for the downlink (base station to subscriber), and single-carrier frequency division multiple access (SC-FDMA) on the uplink (subscriber to base station). OFDMA and SC-FDMA are the state-of-the-art in multiple access technologies, wherein users are assigned separate "subchannels" that effectively divide up the wideband spectrum into a multitude of narrowband spectrum chunks. OFDMA is based on the modulation method called orthogonal frequency division multiplexing

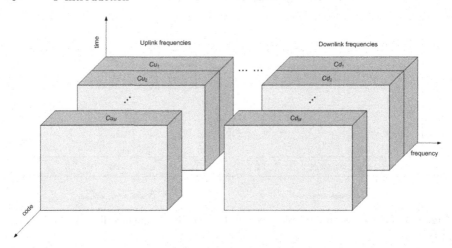

Fig. 1.4. Code division multiple access (CDMA) with frequency division duplexing (FDD).

(OFDM), which similarly uses a multitude of narrowband subcarriers that are orthogonal with each other and carry lower data rate streams, which sum up to a high data rate transmission. We discuss OFDMA and OFDMA in more detail in Sec. 1.2. Details on SC-FDMA can be found in [12], and resource allocation for SC-FDMA have been studied in [13].

Fig. 1.5 shows OFDMA using either TDD or FDD, where a wideband channel is divided up into narrowband subchannels that are orthogonal to each other, such that users can be assigned a mutually exclusive subset of these subchannels without interfering with each other. Both OFDMA and SC-FDMA allow for intelligent scheduling and resource allocation so as to most efficiently use the existing wireless spectrum. The standard assumes a full Internet protocol (IP) network architecture, where the standard voice service is delivered via voice-over-IP (VoIP). 3GPP-LTE is expected to provide:

1. Downlink peak data rates up to 100 Mbps
2. Uplink peak data rates up to 50 Mbps
3. Support for both frequency division duplexing (FDD) and time-division duplexing (TDD)
4. Scalable bandwidths of 1.25, 2.5, 5, 10, 15 and 20 MHz

1.1.2 Evolution of Broadband Access Standards to IEEE 802.16e

Wired Broadband Access

The late 1980s and early 1990s saw a meteoric rise in popularity of the Internet, fueled primarily by the fledgling personal computer, ethernet technology,

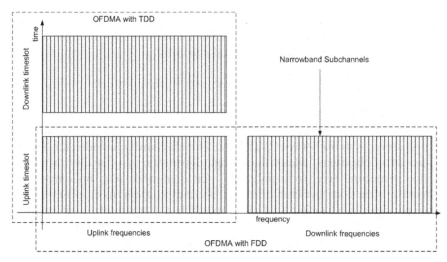

Fig. 1.5. Orthogonal frequency division multiple access (OFDMA) with either TDD or FDD.

and the "killer" application called email. As services and applications over the Internet became more ubiquitous, services that allow business and home consumers to access the high-speed Internet likewise flourished. The dominant technologies for broadband access today are digital subscriber lines and cable modems. Digital subscriber lines, particularly asymmetric digital subscriber lines (ADSL), use discrete multitone (DMT), a multicarrier modulation technique similar to OFDM, to deliver up to 8 Mbps downstream (from network to user) data rates to consumers over telephone lines. Cable modems, on the other hand, use single-carrier 16/64-QAM modulated over a single channel (6 MHz bandwidth), achieving up to 40 Mbps per cable channel.

IEEE 802.16-2001

Although capable of reaching customers in developed urban areas with already existing wired telephone and cable infrastructure, the wired broadband access technologies are unable to reach a lot of suburban and rural customers in a cost-effective manner, particularly those in developing countries. This is because wired infrastructure is typically difficult and expensive to deploy, particularly in areas with rough terrain, e.g. hilly areas [14].

In 2002, the IEEE 802.16-2001 [15] standard was published to provide a common air interface for fixed broadband wireless access systems between 10-66 GHz. The physical layer of IEEE 802.16-2001 uses single-carrier 4, 16, and 64 quadrature amplitude modulation (QAM) and TDMA, and a choice between FDD and TDD. Although spectra is abundant in the 10-66 GHz range of carrier frequencies, the short wavelengths introduce significant deployment challenges, which include strict antenna alignment specifications due to required line-of-sight propagation, and significant attenuations brought about by atmospheric disturbances like rain and snow [14].

IEEE 802.16-2004

In 2004, the IEEE 802.16-2004 [16] standard was published to provide a common air interface for non-line-of-sight (NLOS) operation of fixed broadband wireless access systems between 2-11 GHz. This standard is intended for NLOS residential applications, where line-of-sight operation is typically impractical due to natural (e.g. trees and hills) and man-made (e.g. buildings and bridges) obstructions between the base station to lower rooftop antennas. Three physical layer mechanisms are proposed in the standard: single-carrier, 256-subcarrier OFDM using TDMA, and 2048-subcarrier OFDMA options. Both TDD and FDD options are available in this standard, and peak data rates of up to 23 Mbps in the downlink are possible.

IEEE 802.16e-2005

In 2005, the IEEE 802.16e-2005 [17] standard was published, which extends the IEEE 802.16-2004 standard for combined fixed and mobile broadband wireless access. Focusing primarily on mobility enhancements to IEEE 802.16-2004, this standard similarly supports the three physical layer mechanisms as above. The primary difference is that the OFDMA physical layer in IEEE 802.16e supports varying numbers of subcarriers that scale with the various supported bandwidths, thereby keeping the subcarrier spacing fixed [2] (see Table 1.2). Both TDD and FDD modes are also available in IEEE 802.16e, and peak data rates of 46 Mbps in the downlink and 23 Mbps for the uplink with the 2048-subcarrier, 20 MHz OFDMA physical layer option.

Table 1.2. IEEE 802.16e OFDMA Scalability Parameters [2]

System Bandwidth	Sampling Frequency	FFT Size
1.25 MHz	1.429 MHz	128
2.5 MHz	2.857 MHz	256
5 MHz	5.714 MHz	512
10 MHz	11.429 MHz	1024
20 MHz	22.857 MHz	2048

1.2 Orthogonal Frequency Division Multiple Access

In the previous section, we saw that next-generation wireless standards have embraced OFDMA as the multiple access scheme of choice. In this section, we

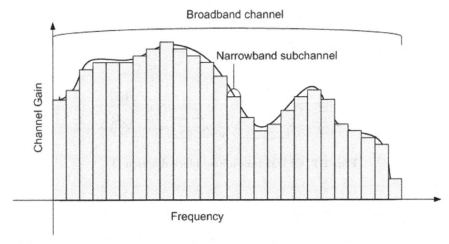

Fig. 1.6. OFDM baseband spectrum, showing the broadband channel subdivided into a multitude of narrowband subchannels.

shall explore the basics of OFDM and OFDMA, and the important problem of resource allocation in OFDMA.

1.2.1 Overview of OFDM

Orthogonal Frequency Division Multiplexing (OFDM) is a multicarrier modulation technique that has been chosen as the modulation scheme for several current and next generation broadband communication systems, e.g. IEEE 802.11a/g wireless local area networks [18], IEEE 802.16-2004/802.16e-2005 wireless metropolitan area networks [16][17], 3GPP-LTE [19], ADSL [20], and power line communications [21]. OFDM is popular especially in broadband wireless communication systems primarily due to its resistance to multipath fading, and its ability to deliver high data rates with reasonable computational complexity. OFDM divides a broadband channel into multiple parallel narrowband subchannels, wherein each subchannel carries low data rate stream, which sums up to a high data rate transmission. A typical OFDM baseband spectrum is shown in Fig. 1.6.

The block diagrams for an uncoded OFDM transmitter and receiver operating over an ideal wireless channel are shown in Figs. 1.7-1.8. The bits are initially mapped by a bank of quadrature amplitude modulation (QAM) encoders into complex symbols, which are then fed into an inverse fast fourier transform (IFFT) to ensure the orthogonality of the subchannels. The output is then converted from parallel to serial and modulated onto a carrier to be transmitted over the air through the wireless channel. At the receiver, the reverse operations are performed. In practical wireless channels, channel

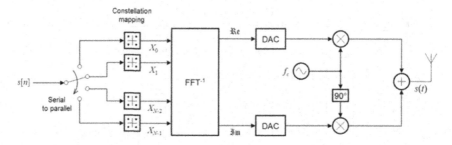

Fig. 1.7. OFDM transmitter block diagram [3]

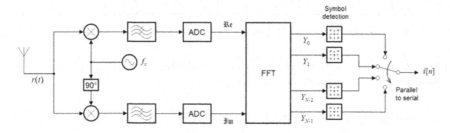

Fig. 1.8. OFDM receiver block diagram [3]

estimation and equalization is necessary to effectively decode the transmitted information.

1.2.2 Overview of OFDMA

In some earlier multi-user wireless systems that used OFDM as the modulation scheme, e.g. IEEE 802.11a/g and IEEE 802.16-2004 OFDM-PHY, a single user is assigned all of the subcarriers at any given instance, and classical TDMA and FDMA is employed to support multiple users. The major setback to this static multiple access scheme is that multiuser diversity is not exploited; i.e. , the fact that the different users that see the wireless channel differently is not being utilized. This led to the development of OFDMA, which allows multiple users to transmit simultaneously on the different subcarriers per OFDM symbol. Since the probability that all users experience a deep fade in a particular subcarrier is typically quite low, intelligent allocation mechanisms can be used to assure that subcarriers are assigned to the users who see "good" channels on them. Fig. 1.9 shows this idea for an OFDMA system with M users that experience different channel gains. This allows the base station, assuming it knows the channel gain information, to allocate resources intelligently in order to maximize some performance metric.

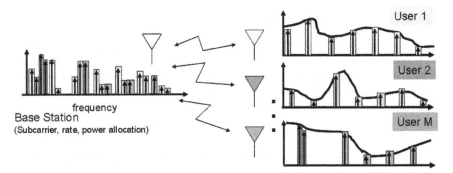

Fig. 1.9. OFDMA resource allocation for M users. Each user is assumed to have statistically independent channel gains, and are allocated a different set of subcarriers by the base station.

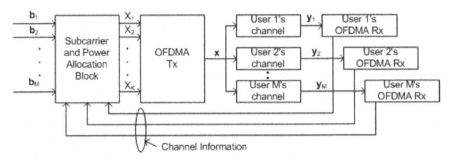

Fig. 1.10. K-subcarrier OFDMA system block diagram for M users. Each user is allocated a different set of subcarriers by the base station.

The block diagram for the downlink of a typical OFDMA system is shown in Fig. 1.10. At the base station transmitter, the bits for each of the different M users $\mathbf{b}_m$ are allocated to the K subcarriers, and each subcarrier k ($1 \leq k \leq K$) of user m ($1 \leq m \leq M$) is assigned a power $p_m(k)$. It is assumed that subcarriers are not shared by different users. Each of the user's bits are then modulated into K L-level QAM symbols X_k, which are subsequently combined using the IFFT into an OFDMA symbol $\mathbf{x}$. This is then transmitted through a time-varying, frequency-selective channel, with each user experiencing an independent channel. The subcarrier allocation is made known to all the users periodically through a control channel; hence, each user needs only to decode the bits on its assigned subcarriers. Note that it is important for the channel state information (CSI) of the users to be known at the transmitter, so that the transmitter can adapt to the time-varying channel conditions, and attempt to use the available resources in the most efficient way.

Resource Allocation in OFDMA

Due to the limited availability of resources at the base station, e.g. bandwidth and power, intelligent allocation of these resources to the users is crucial for delivering the best possible quality of service to the consumer with the least cost. This is especially important with the high data rates envisioned for the next generation wireless standards that utilize OFDMA. The problem of allocating time slots, subcarriers, rates, and power to the different users in an OFDMA system has therefore been an area of active research. Previous research efforts in OFDMA resource allocation in the physical layer have typically focused on the following:

1. ***Formulation****: Maximizing instantaneous performance* Previous research have typically assumed that the allocation decisions are performed only for the current time instant subject to the current resource constraints, and thus have focused only on maximizing the instantaneous performance. Although this reduces the problem into a deterministic optimization problem (which are typically simpler to solve than stochastic optimization problems), the time-varying nature of the wireless channel is not exploited in order to improve the data rate performance of the system.

2. ***Solution****: Developing heuristic sub-optimal algorithms*
   A well-known approach in previous research to achieve near-optimal performance was to relax the exclusive subchannel assignment constraints and solve a large constrained convex optimization problem. Unfortunately, this approach is still too complex for cost-effective real-time implementation. Thus, the focus of previous research has been on developing sub-optimal greedy heuristic algorithms with quadratic complexity and no performance guarantees.

3. ***Assumption****: Assuming perfect channel state information (CSI) available*
   In terms of channel knowledge assumption, previous research typically assumed that the transmitter knows the CSI perfectly at the time the allocation decisions need to be performed. Unfortunately, this assumption is quite unrealistic due to inevitable channel estimation errors and channel feedback delay .

This book attempts to overcome the aforementioned shortcomings, and is summarized in the subsequent section.

1.3 Summary

1.3.1 Thesis Statement

In this book, we defend the following thesis statement

*OFDMA resource allocation problems for instantaneous or ergodic rate al-
location, continuous or discrete rate maximization, with perfect or partial
channel state information assumptions can be solved using dual optimization techniques with linear complexity while achieving negligible optimality
gaps in simulations based on realistic parameters.*

The primary tool that we use to defend this statement is mathematical analysis and optimization theory, supported by practical examples, numerical computations, and Monte-Carlo simulations of OFDMA systems based on the
3GPP-LTE standard.

1.3.2 Contributions

The following is summary of the contributions of this book:

1. ***Formulation****: Maximizing ergodic rates*
   We formulate OFDMA resource allocation problems that maximize the
   ergodic rates instead of instantaneous rates. This allows us to exploit temporal diversity, in addition to frequency and multi-user diversity. It also
   turns out that the computational complexity is even *lower* compared to
   instantaneous performance maximization in practically relevant scenarios
   when using the proposed algorithms.
2. ***Solution****: Developing algorithms based on dual optimization techniques*
   We develop a unified algorithmic framework based on dual optimization
   techniques that is widely applicable to various OFDMA resource allocation problem formulations, e.g. maximizing weighted-sum or proportion-
   ally constrained ergodic or instantaneous rates , considering continuous
   or discrete rates, assuming perfect or partial CSI, and assuming perfect
   or no CDI. It turns out that for most practically relevant formulations,
   the computational complexity can be shown to be linear in the number of
   subcarriers and users, i.e. $\mathcal{O}(MK)$ for an M-user, K-subcarrier OFDMA
   system. Numerical results using 3GPP-LTE OFDMA parameters show
   that the solutions are within 99.9999% of the optimal solution. We also
   develop adaptive algorithms based on *stochastic approximation* principles
   that guarantees convergence with probability one (w.p.1) while significantly decreasing the complexity.
3. ***Assumption****: Assuming that the available CSI is imperfect*
   We consider the scenario when the acquired CSI have errors due to the
   channel estimation and prediction schemes commonly used. Thus, the allocation decisions are made while explicitly considering the error statistics
   of the imperfect CSI. It turns out that neglecting the errors in the CSI
   can result in significant performance degradation.

1.3.3 Organization

This book is organized as follows. Chapter 2 presents a brief survey of previous work with their relative strengths and weaknesses. We also present the OFDMA system model and the key assumptions considered in this book.

Chapter 3 presents downlink OFDMA resource allocation algorithms assuming perfect CSI and perfect channel distribution information (CDI). We consider both continuous (Shannon-capacity) and discrete (Adaptive modulation and coding) ergodic weighted sum-rate maximization with average power constraints. We show that solving this problem using dual optimization techniques involves a single-dimensional line search procedure, wherein each function evaluation in the search procedure involves a single one-dimensional numerical integration, which requires only $\mathcal{O}(MK)$ complexity.

Chapter 4 relaxes the assumption of perfect CSI in Chapter 3 to partial CSI, i.e. wherein only an estimate of the CSI is available. We still assume the knowledge of the distribution information of the partial CSI, and consider the ergodic weighted-sum rate maximization for both continuous and discrete rate cases. We show that the complexity in this case is still $\mathcal{O}(MK)$, but interestingly, due to the availability of closed-form solutions to the expectation integrals, the discrete rate allocation case turns out to be less complex than the continuous rate one.

Chapter 5 presents the OFDMA resource allocation algorithms for ergodic rate maximization with proportional rate constraints. We detail the continuous rate maximization with perfect CSI and CDI case, and show that this problem can be solved by the weighted sum-rate formulation, with optimally chosen user weights. Thus, the technique can be easily extended to the discrete rate and partial CSI cases using the algorithms developed in Chapters 3-4. We also outline an *adaptive* OFDMA resource allocation algorithm based on stochastic approximation methods that simply requires MK operations per symbol *without iterations* and that *do not* require knowledge of the channel distribution information (CDI). We then show that the perfect CDI assumption required in Chapters 3-4 can also be relaxed and solved using this framework.

Finally, Chapter 6 summarizes the contributions of this book, and outlines interesting avenues for future investigation, which include other OFDMA resource allocation formulations, e.g. uplink OFDMA, non-real-time traffic, and outage capacity maximization; resource allocation for OFDMA with multiple transmit and receive antennas (MIMO-OFDMA); multi-cell resource allocation considering inter-cell interference; and multi-hop OFDMA extensions.

1.4 Nomenclature

3GPP-LTE : Third Generation Partnership Project
ADSL   : Asymmetric Digital Subscriber Lines

AMC   : Adaptive Modulation and Coding
AWGN   : Additive White Gaussian Noise
BER   : Bit Error Rate
CDI   : Channel Distribution Information
CDMA   : Code Division Multiple Access
CNR   : Channel-to-noise Raio
CSI   : Channel State Information
DFT   : Discrete Fourier Transform
DMT   : Discrete Multitone
FDD   : Frequency Division Duplexing
FDMA   : Frequency Division Multiple Access
FFT   : Fast Fourier Transform
IEEE   : Institute of Electrical and Electronics Engineers
IFFT   : Inverse Fast Fourier Transform
IID   : Independent and identically distributed
INID   : Independent but not identically distributed
IP   : Internet Protocol
LTE   : Long Term Evolution
MAC   : Media Access Control
Mbps   : Megabits per second
MFI   : Multilevel Fading Inversion
MIMO   : Multiple-input Multiple-output
MWF   : Multilevel waterfilling
NIID   : Not independent but identically distributed
OFDM   : Orthogonal Frequency Division Multiplexing
OFDMA   : Orthogonal Frequency Division Multiple Access
PHY   : Physical Layer
QAM   : Quadrature Amplitude Modulation
QoS   : Quality of Service
SC-FDMA   : Single Carrier - Frequency Division Multiple Access
SNR   : Signal-to-noise Ratio
TDD   : Time Division Duplexing
TDMA   : Time Division Multiple Access
UMTS   : Universal Mobile Telecommunications System
VoIP   : Voice Over Internet Protocol
w.p.1   : With probability one
ZMCSCG   : Zero-mean Circular-symmetric Complex Gaussian

2

Background

2.1 Introduction

In this chapter, we begin by reviewing the seminal and recent work in the field of multi-user wireless communications in Sec. 2.2, with emphasis on physical layer transmit optimization algorithms for OFDMA. This is followed by an exposition of my proposed approach to the problem of OFDMA resource allocation in Sec. 2.3, and a description of the OFDMA system model and key assumptions used throughout this book in Sec. 2.4. Finally, we conclude this chapter in Sec. 2.5.

2.2 Review of Related Work

2.2.1 Scheduling in Wireless Networks

The idea of using channel information at the transmitter to improve the performance of communication systems have been around since at least 1968 [22]. The main concept is to utilize knowledge about the channel to adjust transmission parameters accordingly to maximize communications performance, which is known as *adaptive modulation and coding*. Adaptive modulation and coding in single-user wireless communication systems have been studied extensively (see [23] [24] and the references therein). The extension of the adaptive modulation concept to scheduling in multi-user wireless networks have also been very well studied since the introduction of the concepts of *multiuser diversity* [25] and *proportional fair scheduling* [26]. In these seminal papers, the fading wireless channel was seen as a vehicle to improve the overall system performance when multiple users are involved. The theoretical underpinnings behind this concept, and the fundamental limits of these multiuser channels are addressed by the field of *multiuser information theory*, which is the topic of the next subsection.

2.2.2 Multiuser Information Theory

The focus of this book is on the downlink transmission channel for OFDMA, since this is typically where the increased performance is needed for mobile broadband wireless access applications. This is called a *broadcast channel* [27] in information theory, which consists of a sender with a transmit power and bandwidth budget that is sending independent information simultaneously to multiple users. The capacity and optimal resource allocation for fading broadcast channels has been quite well studied. In [28] and [29], the ergodic and outage capacity , and the optimal resource allocation for a flat-fading broadcast channel was derived. In [30], the capacity region for a frequency-selective broadcast channel with colored Gaussian noise was derived. In [31], the capacity and optimal power allocation for a flat-fading broadcast channel was derived subject to minimum rate constraints. It was shown in the aforementioned publications that superposition coding, followed by successive interference cancellation, is required in order to achieve the capacity of the channel. If we use OFDM transmission with infinitesimally small subcarrier widths to approximate the superposition coding transmission over a frequency-selective channel, some subcarriers would need to be shared among different users, which makes decoding overly complex for practical implementations. Fortunately, the amount of subcarrier sharing is minimal even in the capacity-achieving case [30]. Thus, assigning only one user to each subcarrier could still achieve transmissions close to capacity, and is essentially the downlink OFDMA transmission scheme. However, near capacity performance can be achieved only when optimal allocation of subcarriers, rates, and power is performed.

2.2.3 Physical Layer (PHY) Transmit Optimization

The problem of assigning the subcarriers, rates, time slots, and power to the different users in an OFDMA system has been an area of active research over the past several years. The research in this area can be broadly categorized into two: *margin-adaptive* and *rate-adaptive*. *Margin adaptation* refers to minimizing the transmit power subject to minimum quality of service (QoS) parameters for each user, which could be a combination of data rate, bit error rates, delays, etc. *Rate adaptation* refers to maximizing the data rates subject to various QoS and/or resource constraints.

Margin-adaptive Resource Allocation

In [32], the *margin-adaptive* resource allocation problem was investigated, in which an iterative subcarrier and power allocation algorithm was proposed to minimize the total transmit power given a set of fixed user data rates and bit error rate (BER) requirements. They applied a *constraint relaxation* technique, which allowed the binary integer parameter of subcarrier assignment

to take on real values, which in turn implies a *time-sharing* of each subcarrier among users. This converted the problem into a convex minimization
problem with a convex feasible region, and allowed the use of iterative convex optimization algorithms to find the global minimum transmit power. The
user with the biggest time-sharing factor on each subcarrier is then assigned
to that subcarrier, and a single-user OFDM bit-loading algorithm (see e.g.
[33]) is then run for each user. Although an iterative solution is required in
this algorithm, it is guaranteed to converge to a good solution. Unfortunately,
the algorithm requires a large number of iterations to converge, and is too
complex for cost-effective real-time implementation.

In [34], computationally inexpensive algorithms were proposed to solve
the margin-adaptive problem. They decoupled the problem into a bandwidth
allocation step, which determined the number of subcarriers to be assigned
to each user; and a subcarrier allocation step, which determined the actual
subcarrier assignments to each user. Greedy heuristics were developed for
each of the two steps, and were shown to give comparable performance to the
constraint relaxation technique of [32] with lower complexity.

In [35], an alternative integer programming (IP) formulation, and a linear programming (LP) relaxation algorithm were proposed for the margin-
adaptive problem. It was shown that their methods outperform the constraint
relaxation method in [32] at a lower complexity, but the complexity performance was not justified rigorously. In [36], iterative refinement is used to come
close to the IP solution of [35].

Rate-adaptive Resource Allocation

In [37], the *rate-adaptive* problem was investigated, wherein the objective was
to maximize the total sum continuous rate over all users subject to power and
BER constraints. It was shown in [37] that in order to maximize the total
capacity, each subcarrier should be allocated to the user with the best gain on
it, and the power should be allocated using the water-filling algorithm across
the subcarriers. However, no fairness among the users was considered in [37].
Thus, the users that have the best channel conditions will be assigned all the
resources, which leaves many users without a chance to use the spectrum at all.
The same authors extended the problem formulation to consider *ergodic rates*
in [38], i.e. the expected value of the sum rate is maximized, which utilizes
the temporal dimension when ergodicity of the channel gains is assumed to
improve the data rate performance. However, [38] likewise suffers from the
unfairness problem.

This problem was partially addressed in [39] and [40] by ensuring that each
user would be able to transmit at a minimum rate. The authors of [39] ap-
proached it using two steps similar to [34], wherein the number of subcarriers
and power is initially assigned to each user using a greedy algorithm; followed
by the subcarrier assignment step using the Hungarian algorithm. In [40], the
approach was a simple greedy algorithm that assumes equal power allocation

among subcarriers, and assigned the best subcarrier to each user until the rate requirements for all users are achieved. The remaining subcarriers are then assigned to the users with the best channel gains in them.

In [41], an alternative formulation that maximized the minimum user's data rate was solved by using subcarrier time-sharing methods as in [32]. This enforced a notion of max-min fairness, and thus the starvation of some users in the method of [37] can be avoided. A suboptimal greedy algorithm was also developed which was shown to be close to the relaxed convex problem. This method, though, assumes that all users have similar QoS requirements, which is not the case for practical systems.

In [42], prioritization was enforced using a weighted-sum rate maximization , and a subcarrier time-sharing convex relaxation similar to [32] was used to derive the optimum subcarrier and power allocation. Several greedy algorithms were also proposed to solve the problem with lower complexity. Different weights were assigned to different users, and a higher weight for a user would imply a higher priority of getting resources. By varying the weights for each user's rate, the boundary of the rate-region can also be traced out . In the special case of the weights being identically unity, it would reduce to the problem addressed in [37]. The authors, however, neglected to indicate how the weights are to be assigned in an actual system. More recently, [43] and [44] have discovered a dual optimization framework to solve a similar weighted-sum continuous rate maximization problem. Their work is similar to the approach we advocate in this book, and is one of the special cases that our unified framework can solve (see Section 3.2.6). Note that our contribution in Sec. 3.2.6 was developed independently of [43] and [44].

In [45], the sum data rate was maximized under a *proportional rate constraint*, i.e. the rate of each user should adhere to a set of predetermined proportionality constants. This is a concrete way of assigning priorities to the users, instead of simply assigning arbitrary weights as in [42]. This method is also very useful for service level differentiation, which allows for flexible billing mechanisms for different classes of users. However, the power allocation algorithm proposed in [45] involves solving simultaneous non-linear equations , which requires computationally expensive iterative operations and is thus not suitable for a cost-effective real-time implementation. In cases there the signal-to-noise ratio is high, the algorithm in [45] is shown to reduce to a one-dimensional zero-finding routine, which is much less complex, but may suffer from stability problems. In [46], the strict proportional rate constraints are relaxed to hold approximately, which allowed the power allocation to be solved in closed-form, significantly reducing the complexity, while improving the achieved sum capacity.

Several other methods that use various heuristics have also been proposed. Examples of these include subcarrier partitioning to reduce complexity [47], and game-theoretic Nash bargaining solutions [48].

2.2.4 PHY-MAC Cross-layer Optimization

All of the aforementioned approaches focused on the physical layer transmission optimization for OFDMA. This section reviews several important papers on the PHY-MAC cross-layer approach to OFDMA resource allocation, where longer-term throughput optimality and queue state information is included in their optimization goals.

In [49], resource allocation that optimizes total packet throughput subject to the user's outage probability constraint was proposed. Their algorithm assumes a finite queue size for arrival packets, and dynamically allocates the resources every time-slot based on the users' average SNR, traffic patterns, and QoS requirements. In [50], throughput maximization coupled with queue load balancing was proposed for a simple ON/OFF channel model. Their approach reduced the allocation problem into a maximum weight matching of a bipartite graph, and was shown to stabilize the queues in the OFDMA system, whereas using instantaneous optimization approaches do not.

In [51], an *opportunistic cumulative distribution function (CDF)*-scheduling based subcarrier allocation, and a proportionally-fair power allocation was proposed. Their algorithm was shown to improve overall system capacity in terms of time-average throughput. In [52], a similar opportunistic scheduling algorithm based on [53] that exploits the time varying channel was proposed. In their work, a constant power allocation is assumed, and each user is assigned a *time-slot* for which it could transmit on the assigned subcarrier. Optimal scheduling policies for three QoS/fairness constraints–temporal fairness, utilitarian fairness, and minimum-performance guarantees, were derived to maximize the asymptotic best-case system performance. More recently, in [54] [55], a cross-layer approach that bridges the gap between the physical (PHY) layer and the media access control (MAC) layer was investigated. It was shown that tradeoffs between efficiency and fairness can be realized by maximizing a concave utility function of the user's data rate, instead of maximizing the data rates themselves. Time diversity was also exploited in [55] by maximizing the utility function of an *exponentially weighted* and *time-windowed* average data rate of each user. Prepublished work by the same authors [56] extend the utility based optimization to develop a max-delay-utility scheduling algorithm that utilizes both channel and queue state information.

2.2.5 Comparison of Related Work

Table 2.1 presents a summary of the comparison among several relevant research efforts in OFDMA physical layer transmit optimization. We compare the various research publications in terms of how they formulated the problem, their proposed solution to the problem, and the channel knowledge assumptions that they made. The criteria we use is such that a "Yes" is more desirable in terms of achieving better performance, requiring less computational complexity, or making more realistic assumptions.

Table 2.1. Related work comparison

Criteria / Method	Formulation			Solution		Assumption	
	(1)	(2)	(3)	(4)	(5)	(6)	(7)
Max-min rate [41]	No	No	No	No	No	No	Yes
Sum rate [37][38]	Yes	No	No	Yes	No	No	No
Proportional rate [45][46]	No	No	Yes	No	No	No	Yes
Max-utility [54][55]	[a]No	Yes	Yes	No	No	No	Yes
Weighted rate [43][44]	No	No	Yes	[b]Yes	[c]Yes	No	Yes

[a] Considered some form of temporal diversity by maximizing an exponentially windowed running average of the rate
[b] Independently developed a similar instantaneous continuous rate maximization algorithm
[c] Only for instantaneous continuous rate case, but was not shown in their papers

Criteria

(1) Ergodic rates: The optimization problem is posed such that the *expected value* of the rate is being maximized instead of *instantaneous rate*, which allows the temporal dimension to be exploited when assuming ergodicity of channel gains.

(2) Discrete rates: The practical transmission scheme of only allowing a discrete set of possible data rates is considered rather than just the theoretical continuous rate.

(3) User prioritization: The problem formulation allows setting varying priorities among users to ensure fairness in the system.

(4) Practically optimal: The algorithm is shown in simulations using realistic parameters to have negligible optimality gaps.

(5) Linear complexity: The algorithm can be performed with complexity that is just linear in the number of users and subcarriers.

(6) Imperfect CSI: The algorithm assumes the more realistic scenario of the presence of errors in the available channel state information.

(7) Does not require CDI: The algorithm does not assume knowledge of the probability distribution function of the channel gains, which is difficult to obtain in practice.

In terms of the problem formulation, only [38] considered ergodic rates, and only [55] considered discrete rates. Under the proposed solutions, only [43] and [44] can be considered practically optimal with linear complexity. In terms of channel knowledge assumption, it should be noted that none of the surveyed papers considered imperfect CSI, and only [38] requires CDI since it is also the only work that considers ergodic rate maximization.

2.3 A New Approach to OFDMA Resource Allocation

This book primarily focuses on the physical layer transmit optimization in OFDMA, and assumes that the upper MAC layer performs the other necessary functions, including admission and congestion control, queue management, and user prioritization. This book can thus be seen as a complementary work to the PHY-MAC cross-layer scheduling work, since it extracts further improvements to the physical layer data rate performance in order to benefit the overall system throughput performance.

We observe that in most of the aforementioned work in physical layer transmit optimization, the formulation and algorithms only consider instantaneous performance metrics. Thus, the temporal dimension is not being exploited when the resource allocation is performed. Although the PHY-MAC cross-layer studies performed in [51] and [55] considered time-averaged throughput performance, their channel-based adaptations are based on the average channel-to-noise ratio (CNR), and their approaches focused more on the effect of the past channel information on fairness, rather than exploiting the temporal variations of the wireless channel directly to improve the overall physical data rate performance. We formulate problems considering *ergodic* rates for both continuous (capacity-based) and discrete (Adaptive modulation and coding) rates assuming the availability of the distribution function of the CNR (this assumption is subsequently relaxed in Chapter 5). This allows us to exploit the time dimension explicitly in the formulation, and utilize all three degrees of freedom in our system, namely frequency, time, and multiuser dimensions. Interestingly, when considering ergodic rates, we increase the complexity only slightly during an initialization step, e.g. during frame preamble processing in a frame-based transmission; but actually reduce the complexity when performing the actual resource allocation during data transmission versus instantaneous optimization.

Furthermore, previous research efforts have assumed that algorithms to find the optimal or near-optimal solution to the problem is too computationally complex for real-time implementation. A popular approach to attain near-optimality is *constraint relaxation* (see e.g. [32] [41] [42]). This approach performs a convex reformulation of the problem by relaxing the binary integer constraints $x_{m,k} \in \{0,1\}$ which indicate a subcarrier assignment of user m to subcarrier k; to interval constraints $0 \leq x_{m,k} \leq 1$, where $x_{m,k}$ is now a *sharing factor*. The solution to the reformulated convex problem is then projected back to the original constraint space by assigning each subcarrier to the user with the largest sharing factor. This approach is suboptimal, and more importantly, is also computationally prohibitive, because it involves solving a large constrained convex optimization problem with $2MK$ variables with interval constraints and $K+1$ linear inequality constraints, requiring $\mathcal{O}((2MK)^3)$ operations per iteration when using Newton-type projected gradient methods [57]. Hence, the main focus of previous research have been on developing heuristic approaches with typical complexities in the order of $\mathcal{O}(MK^2)$ (e.g. [34] [42]).

Our approach, on the other hand, is based on a *Lagrangian relaxation* of the power constraints and (possibly) rate constraints, instead of the *constraint relaxation* proposed previously. This relaxation retains the subcarrier assignment exclusivity constraints, but "dualizes" the power/rate constraints and incorporate them into the objective function, thereby allowing us to solve the dual problem instead. This dual optimization framework is much less complex, with complexity order $\mathcal{O}(MK)$; and achieves relative optimality gaps that are less than 10^{-4} (i.e. achieving 99.9999% of the optimal solution) in simulations based on realistic parameters. We also provide adaptive algorithms based on stochastic approximation methods that are shown to converge to the dual optimal solutions w.p.1 with linear complexity *without* the need for iterations. Note that the dual optimization approach is also studied in [43] [44] [58], but their focus has been on instantaneous continuous rate optimization only.

2.4 System Model

In this section, we elaborate on the system model and assumptions considered in this book. Table 2.2 is a notation glossary of the most commonly used terms in this book.

2.4.1 OFDMA Signal Model

We consider a single-cell OFDMA base station, where we ignore the effect of inter-cell interference, which we assume to be either absent (sufficient cell separation given the power budget) or simply modeled as additive white Gaussian noise which increases the noise variance of the signal model. The OFDMA base station has K_{fft} subcarriers with L_{cp} cyclic-prefix, wherein there are K used subcarriers and M active users indexed by the set $\mathcal{K} = \{1, \ldots, k, \ldots, K\}$ and $\mathcal{M} = \{1, \ldots, m, \ldots, M\}$ (typically $K \gg M$) respectively. We assume an average base station transmit power of $\bar{P} > 0$, sampling frequency F_s, bandwidth B, and flat noise power spectral density N_0. The received signal vector for the mth user at the nth OFDM symbol assuming perfect sample and symbol synchronization, and sufficient cyclic prefix length, is given as

$$\boldsymbol{y}_m[n] = \boldsymbol{\Gamma}_m[n]\mathbf{H}_m[n]\boldsymbol{x}_m[n] + \boldsymbol{\nu}_m[n] \tag{2.1}$$

where $\boldsymbol{y}_m[n]$ and $\boldsymbol{x}_m[n]$ are the K-length received and transmitted complex-valued signal vectors; $\boldsymbol{\Gamma}_m[n] = \text{diag}\left\{\sqrt{p_{m,1}[n]}, \ldots, \sqrt{p_{m,K}[n]}\right\}$ is the diagonal gain allocation matrix with $p_{m,k}[n]$ as the power allocated to user m in subcarrier k at time n; $\boldsymbol{\nu}_m[n] \sim \mathcal{CN}(\mathbf{0}, \sigma_\nu^2 \boldsymbol{I}_K)$ with noise variance $\sigma_\nu^2 = N_0 B/K$ is the white zero-mean, circular-symmetric, complex Gaussian (ZMCSCG) noise vector; and

$$\mathbf{H}_m[n] = \text{diag}\{h_{m,1}[n], \ldots, h_{m,K}[n]\} \tag{2.2}$$

is the diagonal channel response matrix.

2.4.2 Multiuser Statistical Fading Channel Model

The diagonal elements $h_{m,k}[n]$ of (2.2) are the complex-valued frequency-domain wireless channel fading random processes for the mth user at the kth

Table 2.2. Notation Glossary

Notation	Description
B	Bandwidth
N_0	Noise power spectral density
F_s	Sampling frequency
N_t	No. of time domain multipath taps
L_{cp}	Length of cyclic prefix
K_{fft}	Number of subcarriers
n	OFDMA symbol index
$h_{m,k}[n]$	Frequency domain complex channel gain
$g_{m,k}[n]$	Time domain complex channel gain
$\mathcal{K}$	Set of used subcarrier indices
K	Number of used subcarriers
k	Subcarrier index
$\mathcal{M}$	Set of active users
M	Number of active users
m	User index
$\mathcal{L}$	Set of discrete rate level indices
L	Number of discrete rate levels
l	Rate level index
r_l	Rate for level l
η_l	SNR upper boundary for rate level l
$\mathcal{L}$	Space of allowable rate vectors
$l_{m,k}$	Rate allocation for user m and subcarrier k
BER_l	Bit error rate for rate level l
$\overline{\text{BER}}$	Average BER constraint
$\mathcal{P}$	Space of allowable power vectors
P	Total power constraint
$p_{m,k}$	Power allocated to user m and subcarrier k
$\gamma_{m,k}$	CNR of user m and subcarrier k
$\hat{\gamma}_{m,k}$	Predicted CNR of user m and subcarrier k
$\gamma_{0,m}$	Cut-off CNR for user m in multi-level waterfilling
σ_ν^2	Ambient noise variance
$\hat{\sigma}_{m,k}^2$	Prediction error variance for user m and subcarrier k
$\rho_{m,k}$	Prediction error to ambient noise ratio
λ	Geometric multiplier
w_m	User weights
$\widehat{}$	Superscript for estimated/predicted terms
$*$	Superscript for optimal terms
$^d/_d$	Superscript/subscript for discrete rate related terms

subcarrier, given as the discrete-time Fourier transform of the N_t time-domain multipath taps $g_{m,i}[n]$ with time-delay τ_i and subcarrier spacing $\Delta f = F_s/K_{\text{fft}}$

$$h_{m,k}[n] = \sum_{i=1}^{N_t} g_{m,i}[n]e^{-j2\pi\tau_i k\Delta f}. \qquad (2.3)$$

The time-domain multipath taps $g_{m,i}[n]$ are modeled as stationary and ergodic discrete-time random processes with normalized temporal autocorrelation function

$$r_{m,i}[\Delta] = \frac{1}{\sigma_{m,i}^2}\mathbb{E}\{g_{m,i}[n]g_{m,i}^*[n+\Delta]\}, \quad i = 1,\ldots,N_t \qquad (2.4)$$

with tap power $\sigma_{m,i}^2$, which we assume to be independent across the fading paths i and across users m. Since $g_{m,i}[n]$ is stationary and ergodic, so is $h_{m,k}[n]$. Hence, the distribution of $\boldsymbol{h}_m[n]$ is independent of n through stationarity, and we can replace time averages with ensemble averages in the problem formulations through ergodicity. In the subsequent discussion, we shall drop the index n when the context is clear for notational brevity.

Although the results in this book are applicable to any stationary fading distribution, we shall prescribe a particular distribution for the fading channels for illustration purposes. We assume that the time domain channel taps are independent ZMCSCG random variables $g_{m,i} \sim \mathcal{CN}(0, \sigma_{m,i}^2)$ with total power $\sigma_m^2 = \sum_{i=1}^{N_t} \sigma_{m,i}^2$. Then from (2.3), we have

$$\boldsymbol{h}_m \sim \mathcal{CN}(\boldsymbol{0}_K, \mathbf{R}_{\boldsymbol{h}_m})$$
$$\mathbf{R}_{\boldsymbol{h}_m} = \mathbf{W}\boldsymbol{\Sigma}_m\mathbf{W}^H \qquad (2.5)$$

where $\mathbf{W}$ is the $K \times N_t$ DFT matrix with entries $[\mathbf{W}]_{k,i} = e^{-j2\pi\tau_i k\Delta f}, k = -K/2 - 1,\ldots,K/2; i = 1,\ldots,N_t$ and $\boldsymbol{\Sigma}_m = \text{diag}\{\sigma_{m,1}^2,\ldots,\sigma_{m,N_t}^2\}$ is an $N_t \times N_t$ diagonal matrix of the time-domain path power[1]. Since we also assume that the fading for each user is independent, then the joint distribution of the stacked fading vector for all users $\boldsymbol{h} = [\boldsymbol{h}_1^T,\ldots,\boldsymbol{h}_M^T]^T$ is likewise a ZMCSCG random vector with distribution $\boldsymbol{h} \sim \mathcal{CN}(\boldsymbol{0}_{KM}, \mathbf{R}_{\boldsymbol{h}})$ where $\mathbf{R}_{\boldsymbol{h}}$ is the $KM \times KM$ block diagonal covariance matrix with $\mathbf{R}_{\boldsymbol{h}_m}$ as the diagonal block elements.

We let $\boldsymbol{\gamma}_m = [\gamma_{m,1},\ldots,\gamma_{m,k}]^T$ where $\gamma_{m,k} = |h_{m,k}|^2/\sigma_\nu^2$ denote the instantaneous channel-to-noise ratio (CNR) with mean $\bar{\gamma}_{m,k} = \sigma_m^2/\sigma_\nu^2$. Note that $\gamma_{m,k}$ for a particular subcarrier k and different users m are independent but not necessarily identically distributed (INID) exponential random variables; and for a particular user m and different subcarriers k are not independent but identically distributed (NIID) exponential random variables.

---

[1] Following the convention in [17] and [19], we assume that the number of used subcarriers K is odd by including the null subcarrier at index 0 as part of the used subcarriers.

2.4.3 Optimization Variables

Denote by $\boldsymbol{p} = [\boldsymbol{p}_1^T, \cdots, \boldsymbol{p}_K^T]^T$ the length MK vector of power allocation values to be determined, where $\boldsymbol{p}_k = [p_{1,k}, \cdots, p_{M,k}]^T$ is the M-length vector of power allocation values with $p_{m,k}$ as the assigned power for user m in a subcarrier k. Although subcarrier, rate, and time slot allocation is required, in addition to determining the power values, it can be seen that the power vector can essentially capture these other resource assignments as well.

Subcarrier Allocation

The exclusive subcarrier allocation restriction in OFDMA can be captured by constraining the power vector as $\boldsymbol{p}_k \in \mathcal{P}_k \subset \mathbb{R}_+^M$, where the space of allowable power vectors is

$$\mathcal{P}_k \equiv \{\boldsymbol{p}_k \in \mathbb{R}_+^M | p_{m,k} p_{m',k} = 0; \forall m \neq m'; m, m' \in \mathcal{M}\} \quad (2.6)$$

For notational convenience, we let $\boldsymbol{p} \in \mathcal{P} \equiv \mathcal{P}_1 \times \cdots \times \mathcal{P}_K \subset \mathbb{R}_+^{MK}$ denote the space of allowable power vectors for all subcarriers.

Continuous Rate Allocation

The continuous rate or capacity for user m and subcarrier k is given as

$$R_{m,k}(p_{m,k}\gamma_{m,k}) = \log_2(1 + p_{m,k}\gamma_{m,k}) \quad \text{bps/Hz} \quad (2.7)$$

Thus, the power allocation value $p_{m,k}$ determines a unique rate allocation, and $p_{m,k} = 0$ also results in zero rate allocation, which of course also means that the subcarrier k is not assigned to user m.

Discrete Rate Allocation

In the discrete rate allocation case, the data rate of the kth subcarrier for the mth user can be given by the staircase function

$$R_{m,k}^d(p_{m,k}\gamma_{m,k}) = \begin{cases} r_0, & \eta_0 \leq p_{m,k}\gamma_{m,k} < \eta_1 \\ r_1, & \eta_1 \leq p_{m,k}\gamma_{m,k} < \eta_2 \\ \vdots, & \vdots \\ r_{L-1}, & \eta_{L-1} \leq p_{m,k}\gamma_{m,k} < \eta_L \end{cases} \quad (2.8)$$

where $\{\eta_l\}_{l \in \mathcal{L}}$, $\mathcal{L} = \{0, \ldots, L-1\}$, are the SNR boundaries which define a particular code-rate and constellation pair combination that result in r_l data bits per transmission with a predefined target bit error rate (BER), and where $r_l \geq 0$, $r_{l+1} > r_l$, $r_0 = 0$, $\eta_0 = 0$, and $\eta_L = \infty$. Thus, similar to the continuous rate case, the power allocation value $p_{m,k}$ determines a unique rate allocation

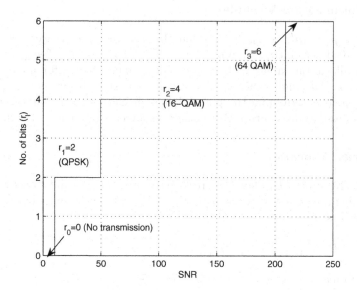

Fig. 2.1. Example discrete-rate function for an uncoded system with BER=10^{-3}. Note that the SNR is plotted in linear and not dB scale.

for a particular target BER, and $p_{m,k} = 0$ also results in zero rate allocation. We assume a Grey-coded square 2^{r_l}-QAM modulation scheme, where the BER without channel coding in AWGN can be approximated to within 1-dB for $r_l \geq 2$ and BER $\leq 10^{-3}$ by BER $\approx 0.2e^{\left[\frac{-1.6 p_{m,k}\gamma_{m,k}}{2^{r_l}-1}\right]}$ [24]. Fig. 2.1 shows an example of a discrete rate function for rate levels $r_l = \{0, 2, 4, 6\}$ corresponding to no transmission, QPSK, 16-QAM, and 64-QAM transmission, and SNR boundaries $\eta_l \in \{0, 9.93, 49.66, 208.45\}$ with a BER constraint of 10^{-3}.

Time slot allocation

In the context of OFDMA, a time slot can be considered as a single OFDMA symbol (or several OFDMA symbols), and time slot allocation in this case is more granular than conventional TDMA time slot allocation since each OFDMA symbol may be shared by more than a single user. Hence, time slot allocation fundamentally entails performing the OFDMA resource allocation algorithms across time for each OFDMA symbol. In the previous work that considered instantaneous rate allocation only, the OFDMA algorithms were simply re-run every symbol (or several symbols). In this book, we can capture the idea of "time slot allocation" by using the ergodicity assumption , and determine *power allocation functions* that are parameterized by the channel knowledge. For example, if we assume perfect channel knowledge, then our optimization variable is essentially

$$\boldsymbol{p}(\cdot) \in \boldsymbol{\mathcal{P}} \equiv \left\{ \mathbb{R}_+^{MK} \to \mathbb{R}_+^{MK} : p_{m,k}(\cdot)p_{m',k}(\cdot) = 0 \text{ w.p.1}, \forall m' \neq m \right\} \qquad (2.9)$$

whose search space includes all $\mathbb{R}_+^{MK}$-measurable functions with exclusive sub-carrier allocation restriction imposed w.p.1. In the case of the adaptive algorithms discussed in Chapter 5, the power allocation is indexed by the time index n, i.e. $\boldsymbol{p}[n]$ and the exclusive subcarrier allocation restriction is simply imposed as $p_{m,k}[n]p_{m',k}[n] = 0, \forall m' \neq m, \forall n$.

2.4.4 PHY-MAC Interaction

The resource allocation problems considered in this book include assigning the power, subcarriers, rates, and time slots to the different users such that weighted-sum rate (Chapters 3-4) or sum rate subject to proportional rate constraints (Chapter 5) of the users are maximized. Although the focus of this book is primarily on the physical layer transmit optimization, it is important to discuss our assumptions on the cross-layer PHY-MAC interactions in order to see how one can apply the results in PHY-MAC cross-layer optimization discussed in Sec. 2.2.4. Specifically, we assume that the upper MAC layer passes the following information to the physical layer optimization routine:

- Set of active users $\mathcal{M}$: The MAC layer performs the necessary admission and congestion control to determine which are the active users at a particular time
- Priority for the active users w_m or ϕ_m for all $m \in \mathcal{M}$: Depending on queue back-logs and information on the average data rate for each user, the MAC layer sets the appropriate user weights w_m in the weighted-sum rate maximization formulations, or the user proportionality values ϕ_m in the proportional rate formulations.

There are numerous ways in which the MAC layer can determine these parameters, but are beyond the intended scope of this book. Admission and congestion control to determine the active user set depending on the utility of the network and availability of the resources are studied in [55] [59]. User prioritization by setting the weights w_m as the reciprocal of the user's average rate so far has been shown to approximate proportional fairness [55]. Another possibility is to set the weights as a directly proportional function of the queue-back log of the user, which can be shown to minimize the delay and ensure network stability [56].

2.5 Conclusion

In this chapter, we surveyed several important papers in OFDMA resource allocation, and showed the relative strengths and weaknesses of each of these. We then presented the general idea of our new approach to OFDMA resource allocation based on dual optimization techniques. We also presented the system model and key assumptions used in this book.

Chapters 3-4 shall elaborate on the dual optimization framework for solving the weighted-sum rate maximization problem in OFDMA with channel distribution information, where we assume perfect and partial channel state information, respectively. Chapter 5 presents an extension of the framework to formulations that have proportional rate constraints with or without channel distribution information. Chapter 6 then concludes this book.

3

Weighted-sum rate Maximization with Perfect CSI

3.1 Introduction

In this chapter, we consider the weighted-sum rate maximization problem subject to a single average power constraint by assuming the availability of perfect CSI and CDI. This is most suitable to downlink OFDMA with best-effort traffic, wherein the user weights can be used to enforce certain notions of fairness (e.g. proportional fairness can be attained by setting the user weights as the reciprocal of the user's average rate so far [55]). We formulate the problem considering *ergodic* rates for both continuous (capacity-based) and discrete (adaptive modulation and coding) rates. Note that a similar ergodic formulation for continuous rate maximization has been considered in [38], but they limited their study to the case of maximizing the unweighted sum capacity, and they did not propose efficient algorithms to solve the problem. The contents of this chapter are close to that of the papers [60] [61] [62].

This chapter is organized as follows. Section 3.2 focuses on the continuous rate case, which is equivalent to a *Shannon-capacity* based formulation. The results of this section have more theoretical than practical value, and elaborates on the details of the dual optimization approach to OFDMA resource allocation. Section 3.3 focuses on the discrete rate case, which is equivalent to a more practical *adaptive modulation and coding* (AMC) scenario. Section 3.4 presents numerical results to corroborate our analysis, and we conclude this chapter in Section 3.5.

3.2 Continuous Rate Maximization with perfect CSI and CDI

3.2.1 Problem Formulation

Since we assume perfect CSI, we consider the optimization variable $\boldsymbol{p}(\cdot)$ in (2.9) as a function of the realization of the fading CNR of all users

$\gamma = [\gamma_1^T, \ldots, \gamma_M^T]^T$. We also assume that we have perfect channel distribution information (CDI), i.e. we know the stationary pdf of γ, thereby allowing us to take the expectation . The ergodic weighted sum capacity maximization problem is then

$$f^* = \max_{p(\cdot) \in \mathcal{P}} \ \mathbb{E}_\gamma \left\{ \sum_{m \in \mathcal{M}} w_m \sum_{k \in \mathcal{K}} R_{m,k} \left(p_{m,k} \gamma_{m,k} \right) \right\}$$

$$\text{s.t.} \ \ \mathbb{E}_\gamma \left\{ \sum_{m \in \mathcal{M}} \sum_{k \in \mathcal{K}} p_{m,k} \right\} \le \bar{P}$$

$$(3.1)$$

where $R_{m,k}$ is given in (2.7) and $\mathcal{P}$ is given in (2.9).

Comments on the user weighted formulation

The user weights w_m in (3.1) are positive constants such that $\sum_{m \in \mathcal{M}} w_m = 1$. Theoretically, varying these weights allows us to trace out the ergodic capacity region [28]; algorithmically, varying the weights allows us to prioritize the different users in the system and enforce certain notions of fairness. Note that the choice of w_m is typically handed down to the physical layer from a higher layer, e.g. the MAC layer. A possible choice is $w_m[n] = 1/R_m[n]$ where $R_m[n]$ is the average rate for user m so far at time n, which was shown to approximate proportional fairness [55].

A caveat for this ergodic weighted sum capacity formulation, however, is that the w_m terms need to be held constant for a time period that allows the ergodicity property of the channels gains to kick in, which may hurt the fairness of the system. Fortunately, in next generation OFDMA implementations (e.g. IEEE 802.16e [17] and 3GPP-LTE [19]), the MAC layer hands down user-weights to the physical layer on a per-frame (or longer) basis. This is because holding weights constant for a period is beneficial from a system implementation complexity perspective, thereby requiring less signaling and feedback overhead , while still enforcing fairness, albeit on a larger timescale. Thus, depending on the frame length (which in IEEE 802.16e can reach up to 20ms [17]) and the mobile speed, ergodicity can be assumed in a lot of cases within the frame, and the ergodic weighted sum capacity formulation is ideal in these scenarios. A comparison of the fairness in ergodic and instantaneous rate formulations, and the effect of different w_m terms on overall communication performance of the system, however, is beyond the intended scope of this book.

Comments on the average power constraint

By enforcing the average power constraint in (3.1), we would like to keep a handle on the average power at the base station transmitter, in order to conserve power and more importantly, to prevent overheating. However, this constraint allows instantaneous power levels to exceed the average power when necessary . Since practical power amplifiers have a limited linear region, then a peak power constraint is likewise important. Although we do not include this constraint for simplicity of presentation, it can be shown that algorithms similar to the ones proposed in this book can be easily modified to impose this constraint (see [13] for an example of this extension).

Problem Classification

The problem in (3.1) is part of a class of optimization problems called *infinite dimensional stochastic programs*. The stochastic program in this case is further classified as an *adaptive* or *anticipative* model [63, Sec. 1.4], i.e. we are allowed to make an observation, the realization of γ, before making our decision on the power allocation vector $p(\cdot)$. Thus, we actually allow a large class of solutions, i.e. $\mathbb{R}_+^{MK}$-measurable functions, subject to the exclusive subcarrier allocation constraint *on each possible realization* of γ, which is defined in (2.9). Fortunately, familiar concepts in deterministic optimization, e.g. duality, are founded on general geometrical concepts, and are thus also applicable to this *infinite dimensional* space [64]. Thus, using variational calculus techniques [65], we can extend concepts familiar to deterministic optimization like gradients, subgradients, and Lagrangian duality to this infinite dimensional space.

3.2.2 Dual Optimization Framework

Note that the objective function in (3.1) is concave, but the constraint space $\mathcal{P}$ is highly non-convex (it is in fact a discrete space), and is in general very difficult to solve. Fortunately, (3.1) is separable across the subcarriers , and is tied together only by the power constraint. In these problems, it is useful to approach the problem using duality principles [58] [57]. Let us write the Lagrangian

$$L(p(\cdot),\lambda) = \mathbb{E}_\gamma \left\{ \sum_{m\in\mathcal{M}} w_m \sum_{k\in\mathcal{K}} R_{m,k}\left(p_{m,k}\gamma_{m,k}\right) \right\} + \lambda \left(\bar{P} - \mathbb{E}_\gamma \left\{ \sum_{k\in\mathcal{K}} \sum_{m\in\mathcal{M}} p_{m,k} \right\} \right) \tag{3.2}$$

The dual problem is defined as

$$g^* = \min_{\lambda \geq 0} \Theta(\lambda) \tag{3.3}$$

where the dual objective is given by

$$\Theta(\lambda) = \max_{\boldsymbol{p}(\cdot)\in\mathcal{P}} L(\boldsymbol{p}(\cdot),\lambda) \tag{3.4a}$$

$$= \lambda\bar{P} + \max_{\boldsymbol{p}(\cdot)\in\mathcal{P}} \sum_{k\in\mathcal{K}} \mathbb{E}_{\boldsymbol{\gamma}} \left\{ \sum_{m\in\mathcal{M}} \left(w_m R_{m,k}\left(p_{m,k}\gamma_{m,k}\right) - \lambda p_{m,k}\right) \right\} \tag{3.4b}$$

$$= \lambda\bar{P} + \sum_{k\in\mathcal{K}} \max_{\boldsymbol{p}_k(\cdot)\in\mathcal{P}_k} \mathbb{E}_{\boldsymbol{\gamma}} \left\{ \sum_{m\in\mathcal{M}} \left(w_m R_{m,k}\left(p_{m,k}\gamma_{m,k}\right) - \lambda p_{m,k}\right) \right\} \tag{3.4c}$$

$$= \lambda\bar{P} + \sum_{k\in\mathcal{K}} \mathbb{E}_{\boldsymbol{\gamma}} \left\{ \max_{\boldsymbol{p}_k(\cdot)\in\mathcal{P}_k} \sum_{m\in\mathcal{M}} \left(w_m R_{m,k}\left(p_{m,k}\gamma_{m,k}\right) - \lambda p_{m,k}\right) \right\} \tag{3.4d}$$

$$= \lambda\bar{P} + K\mathbb{E}_{\boldsymbol{\gamma}_k} \left\{ \max_{m\in\mathcal{M}} \left[\max_{p_{m,k}\geq 0} \left(w_m R_{m,k}\left(p_{m,k}\gamma_{m,k}\right) - \lambda p_{m,k}\right) \right] \right\} \tag{3.4e}$$

where (3.4a) is the dual objective; (3.4b) follows from the linearity of the expected value ; (3.4c) follows from the fact that the power variables are separable across the subcarriers[1]; (3.4d) follows from the fact that the power variables are a function of each realization of $\boldsymbol{\gamma}$, thereby allowing us to interchange the order of maximization and expected value; and (3.4e) follows from the exclusive subcarrier assignment constraint and the fact that the channel gains are NIID across subcarriers. Note that we have reduced the problem to a per-subcarrier optimization, and since $K \gg M$, we have significantly decreased the computational burden.

The innermost maximization between the square brackets in (3.4e) has a simple closed-form expression for the optimal power given as

$$\tilde{p}_{m,k}(\lambda) = \left[\frac{1}{\gamma_{0,m}(\lambda)} - \frac{1}{\gamma_{m,k}} \right]^+ \tag{3.5}$$

where $[x]^+ = \max(0, x)$ and $\gamma_{0,m}(\lambda) = \frac{\lambda\ln 2}{w_m}$, which is a simple "multi-level water-filling" power allocation with cut-off CNR $\gamma_{0,m}(\lambda)$, below which we do not transmit any power, and above which we transmit more power when the CNR $\gamma_{m,k}$ is higher.

Using (3.5) in (3.4e), the dual problem in (3.3) can be written as

$$g^* = \min_{\lambda\geq 0} \left[\lambda\bar{P} + K\mathbb{E}_{\boldsymbol{\gamma}_k} \left\{ g_k(\boldsymbol{\gamma}_k, \lambda) \right\} \right] \tag{3.6}$$

$$g_k(\boldsymbol{\gamma}_k, \lambda) = \max_{m\in\mathcal{M}} \left\{ g_{m,k}(\gamma_{m,k}, \lambda) \right\} \tag{3.7}$$

where (3.7) is a max function over the M per-subcarrier *marginal dual* functions

---

[1] The separability is due to the fact that the exclusive subcarrier allocation constraint is enforced on a per-subcarrier basis (see (2.6)), and that the average power constraint that ties the power variables across subcarriers has been "dualized" into the Lagrangian objective function

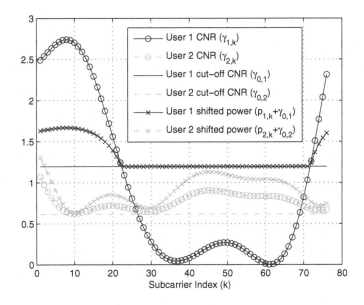

Fig. 3.1. Multi-level waterfilling snapshot for a 2-user, 76-subcarrier system.

$$g_{m,k}(\gamma_{m,k}, \lambda) = w_m R_{m,k}\left(\tilde{p}_{m,k}(\lambda)\gamma_{m,k}\right) - \lambda\tilde{p}_{m,k}(\lambda)$$
$$= \left(\frac{w_m}{\ln 2} \ln\left(\frac{\gamma_{m,k}}{\gamma_{0,m}(\lambda)}\right) - \frac{w_m}{\ln 2} + \frac{\lambda}{\gamma_{m,k}}\right) u(\gamma_{m,k} - \gamma_{0,m}(\lambda))$$

(3.8)

and where

$$u(x) = \begin{cases} 0, & x < 0 \\ 1, & x \geq 0 \end{cases}$$

is the unit (Heaviside) step function. Observe that the dual-optimal subcarrier allocation policy is to assign the subcarrier to the user with the maximum *marginal dual*, and we call this the "max-dual user selection." Note also that (3.8) is non-negative and is not differentiable at $g_{m,k}(\gamma_{0,m}(\lambda), \lambda) = 0$.

Fig. 3.1 shows an instantaneous snapshot of the multi-level waterfilling power allocation for a 2-user, 76-subcarrier system with user weights $w = [0.34, 0.66]$. We show the CNR values $\gamma_{m,k}$, the cut-off CNR $\gamma_{0,m}$, and the multi-level waterfilling power allocation shifted by the cut-off for illustration purposes $\tilde{p}_{m,k} + \gamma_{0,m}$. We see that no power is allocated to CNRs below the cut-off, and the higher CNRs get higher power. Fig. 3.2 shows the resulting instantaneous marginal duals $g_{m,k}$, and the corresponding optimal power allocation $p^*_{m,k}$ for the current instance. We see that the subcarrier is allocated to the user with higher marginal dual value.

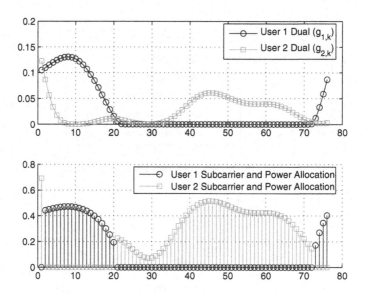

Fig. 3.2. Per-user and per-subcarrier dual $g_{m,k}$ and the corresponding optimal subcarrier and power allocation $p^*_{m,k}$.

3.2.3 Numerical Evaluation of the Expected Dual

Computing the expectation in (3.6) in a straightforward manner involves an $M-$ dimensional integral over the joint pdf of the $M-$length fading vector γ_k, which is typically too complex to solve using direct numerical integration techniques (e.g. Gaussian quadrature) except for small M, e.g. 2 or 3, since this requires $\mathcal{O}(N^M)$ computations where N is the number of function evaluations required for a one-dimensional integral with the same accuracy [66]. However, if we can somehow compute a closed-form expression for the pdf of (3.7), then we can reduce the expectation to just a one-dimensional integral that is solvable in $\mathcal{O}(MN)$. Since $\gamma_{m,k}$ for different ms are INID, then (3.8) is likewise INID for different ms. Thus, (3.7) is the largest order statistic of INID random variables $g_{m,k}(\gamma_{m,k}, \lambda)$ with pdf [67, Sec. 5.2]

$$f_{g_k}(g_k) = \prod_{m \in \mathcal{M}} F_{g_{m,k}}(g_k) \left(\sum_{m \in \mathcal{M}} \frac{f_{g_{m,k}}(g_k)}{F_{g_{m,k}}(g_k)} \right) \qquad (3.9)$$

where $F_{g_{m,k}}(g_{m,k})$ and $f_{g_{m,k}}(g_{m,k})$ are the cumulative distribution function (CDF) and probability density function (PDF) of $g_{m,k}(\gamma_{m,k}, \lambda)$, respectively.

In order to derive these distribution functions given the distribution $F_{\gamma_{m,k}}(\gamma_{m,k})$ of $\gamma_{m,k}$, we need an expression for the inverse function of $g_{m,k}(\gamma_{m,k}, \lambda)$, which is given as (see Appendix A)

$$\breve{\gamma}_{m,k}(g_{m,k}) = \frac{-\gamma_{0,m}(\lambda)}{W\left(-e^{\left(-g_{m,k}\frac{\ln 2}{w_m}-1\right)}\right)} u(g_{m,k}) \qquad (3.10)$$

where $W(x)$ is the *Lambert-W* function, which is the solution to the transcendental equation $W(x)e^{W(x)} = x$. This function is ubiquitous in the physical sciences, and efficient algorithms have been developed for its computation [68]. Note that $\breve{\gamma}_{m,k}(0) = \gamma_{0,m}(\lambda)$ as expected.

Using this expression for the root, we can then derive the cdf of $\gamma_{m,k}$ as [69]

$$F_{g_{m,k}}(g_{m,k}) = F_{\gamma_{m,k}}(\breve{\gamma}_{m,k}(g_{m,k}))\, u(g_{m,k}) \qquad (3.11)$$

The pdf is then given as the derivative of (3.11) with respect to $g_{m,k}$

$$f_{g_{m,k}}(g_{m,k}) = F_{\gamma_{m,k}}(\gamma_{0,m}(\lambda))\,\delta(g_{m,k}) + f_{\gamma_{m,k}}(\breve{\gamma}_{m,k}(g_{m,k}))\frac{\breve{\gamma}_{m,k}^2(g_{m,k})}{\breve{\gamma}_{m,k}(g_{m,k})\frac{w_m}{\ln 2}-\lambda}u(g_{m,k}) \qquad (3.12)$$

where $\delta(x)$ is the Dirac delta functional[2]. Finally, using (3.11) and (3.12) in (3.9) and then in (3.6), our dual problem can now be written as

$$g^* = \min_{\lambda \geq 0}\left[\lambda\bar{P} + K\int_0^\infty g_k f_{g_k}(g_k)dg_k\right] \qquad (3.13)$$

3.2.4 Optimal Subcarrier and Power Allocation

Using standard duality arguments (see e.g. [57, Prop. 5.1.2]), the dual objective function in (3.13) can be shown to be convex in the single variable λ, and is therefore unimodal [57, App C.3]. Thus, we can use derivative-free line search procedures, e.g. Golden-section or Fibonacci search [66] to find the optimal λ^*. In our numerical experiments using the fminbnd[3] function in Matlab, we achieve convergence for typical wireless scenarios within a tolerance of 10^{-4} in less than 10 iterations.

Once we determine λ^*, we plug it back into the optimal power allocation function and arrive at the following simple user assignment and power allocation for each subcarrier k given as

$$m_k^* = \arg\max_{m\in\mathcal{M}}\{w_m R_{m,k}(\tilde{p}_{m,k}(\lambda^*)\gamma_{m,k}) - \lambda^*\tilde{p}_{m,k}(\lambda^*)\} \qquad (3.14)$$

$$p_{m,k}^* = \tilde{p}_{m,k}(\lambda^*)\mathbf{1}(m = m_k^*) \qquad (3.15)$$

where $\mathbf{1}(x)$ is the indicator function, which evaluates to 1 if x is true and 0 if false. Fig. 3.3 presents a flow chart of the algorithm.

---

[2] Note that $F_{g_{m,k}}(g_{m,k})$ is discontinuous at $g_{m,k} = 0$ with $F_{g_{m,k}}(0^-) = 0$ and $F_{g_{m,k}}(0^+) = F_{\gamma_{m,k}}(\gamma_{0,m}(\lambda))$.

[3] fminbnd uses a combination of Golden-section search and parabolic interpolation.

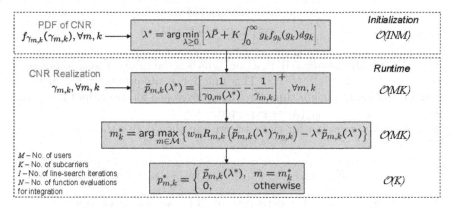

Fig. 3.3. OFDMA resource allocation algorithm for ergodic weighted-sum continuous rate maximization.

Note that it is possible that the dual optimal power allocation values do not satisfy the total power constraint. Hence, our final power allocation values should be multiplied by $\eta = \bar{P}/P_{tot}(\lambda^*)$ where

$$\hat{P}_{tot}(\lambda^*) = \mathbb{E}_{\gamma}\left\{ \sum_{m\in\mathcal{M}} \sum_{k\in\mathcal{K}} p_{m,k}^* \right\} \qquad (3.16)$$

which we plug back into the objective in (3.1) to arrive at our computed primal optimal value

$$\hat{f}^* = \mathbb{E}_{\gamma}\left\{ \sum_{m\in\mathcal{M}} w_m \sum_{k\in\mathcal{K}} \log_2(1 + \eta\gamma_{m,k}p_{m,k}^*) \right\} \qquad (3.17)$$

3.2.5 Complexity Analysis

Once we determine λ^* by solving (3.13), we do not need to update it as long as the statistics of the fading channel vector γ remain the same. Thus, the complexity of resource allocation requires an initial $\mathcal{O}(INM)$ computations to determine λ^*, where I is the number of iterations for the line search procedure to converge, and N is the number of function evaluations to compute the dual objective integral. The allocation in (3.14)-(3.15) needs $\mathcal{O}(MK)$ computations per symbol.

3.2.6 Instantaneous Weighted Sum Rate Maximization

Although we have focused on the ergodic rate maximization problem, our duality framework can be simplified to solve the instantaneous rate maximization problem given as

$$f_{\text{inst}}^* = \max_{\boldsymbol{p} \in \mathcal{P}} \quad \sum_{m \in \mathcal{M}} w_m \sum_{k \in \mathcal{K}} R_{m,k}\left(p_{m,k}\gamma_{m,k}\right)$$

$$\text{s.t.} \quad \sum_{m \in \mathcal{M}} \sum_{k \in \mathcal{K}} p_{m,k} \leq \bar{P} \tag{3.18}$$

and is essentially identical to the problem considered in [42], which was solved using a convex relaxation of the above problem by relaxing the exclusive subcarrier assignment constraint to one where subcarrier sharing is allowed through a sharing factor.

We use the dual optimization approach, where the dual problem can be derived similarly as the ergodic case

$$g_{\text{inst}}^* = \min_{\lambda \geq 0} \left[\lambda \bar{P} + \sum_{k \in \mathcal{K}} \max_{m \in \mathcal{M}} \left\{ w_m R_{m,k}\left(\tilde{p}_{m,k}(\lambda)\gamma_{m,k}\right) - \lambda \tilde{p}_{m,k}(\lambda) \right\} \right] \tag{3.19}$$

where $\tilde{p}_{m,k}(\lambda)$ is the same power allocation function given in (3.5). Note the similarity to the ergodic case (c.f. (3.3)-(3.4e)), where the primary difference is that the expected values are no longer present. Using a similar line search procedure, we can find the optimal λ_{inst}^* and end up with the same optimal subcarrier and power allocation functions as in (3.14)-(3.15). One subtle, albeit important difference, is that in instantaneous maximization, the optimal λ_{inst}^* is dependent on each channel realization γ, and thus needs to be computed every time the channel changes. This is in contrast to the ergodic maximization case where the λ^* depends on the distribution function of the channel $f_{\gamma}(\gamma)$, and thus needs to be computed only when the statistics of the channel have changed. Thus, although the initialization for the ergodic maximization is more complex, the per-symbol resource allocation complexity ends up being lower than the instantaneous optimization case. Furthermore, because the total power in each time instant is constrained to be less than or equal to $\bar{P}$ in the instantaneous case, there is no flexibility of allowing the total power in each time instant to vary (while still maintaining the average power constraint across time) unlike the ergodic maximization case. Fig. 3.4 presents a flow chart of the OFDMA instantaneous weighted-sum continuous rate maximization algorithm.

Searching for λ_{inst}^*

It is important to point out that [43] and [44] have independently come up with an identical "multi-level waterfilling" power allocation with "max-dual user selection". However, they proposed to compute the optimal λ_{inst}^* using a single-dimensional subgradient search (see Sec. 5.2.2 for a description of the subgradient search). Since the subgradient search is iterative in nature and cannot terminate in a fixed number of iterations, its implementation is potentially difficult. In this section, we propose an efficient method to find λ_{inst}^* using line-search techniques which requires a predictable number of iterations, and is thus suitable for hardware implementation.

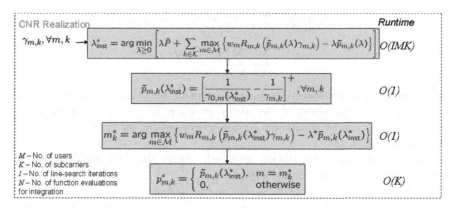

Fig. 3.4. OFDMA resource allocation algorithm for instantaneous weighted-sum continuous rate maximization.

In derivative-free line search techniques, e.g. Golden section or Fibonacci search, once an interval is determined where the optimal solution definitely lies, i.e. $\lambda_{\text{inst}}^* \in [\lambda_{min}, \lambda_{max}]$, termination is guaranteed in

$$I = \left\lceil \frac{\ln\left(\frac{\epsilon}{\lambda_{max} - \lambda_{min}}\right)}{\log(0.618)} + 1 \right\rceil$$

iterations, where ϵ is the desired tolerance [66]. The following proposition establishes an easily computable interval over which we can search for λ_{inst}^*.

Proposition 3.1. *Suppose λ_{inst}^* satisfies the total power constraint tightly, i.e.* $\sum\limits_{k \in \mathcal{K}} p_{m_k^*, k}^* = \bar{P}$, *then*

$$\lambda_{min} = \frac{\ln 2}{K \min\limits_{m} w_m}\left(\bar{P} + \sum_{k \in \mathcal{K}} \max_{m} \frac{1}{\gamma_{m,k}}\right) \leq \lambda_{\text{inst}}^* \leq \frac{K}{\bar{P} \ln 2} \max_{m \in \mathcal{M}} w_m = \lambda_{max}$$

(3.20)

Proof. See Appendix B.

Note that computing the lower bound requires $\mathcal{O}(MK)$ operations and the upper bound requires $\mathcal{O}(M)$ operations, which do not change the overall complexity order. One caveat in using the interval specified in (3.20) is that the assumption of λ_{inst}^* satisfying the total power constraint only holds approximately. However, as long as K is large, the constraint holds tightly (see Sec. 3.2.8), and coupled by the bounds being conservative, one should not worry about the λ_{inst}^* not being bracketed by these values.

3.2.7 constant power Allocation

It has been established in previous research that constant power allocation actually performs as well as optimal water-filling, esp. in high SNR cases [24]. Under the constant power allocation assumption, the power is set to $\bar{P}/K$, and the subcarrier allocation is simplified to

$$m_k^* = \arg\max_{m \in \mathcal{M}} \left\{ w_m R_{m,k} \left(\frac{\bar{P}}{K} \gamma_{m,k} \right) \right\} \tag{3.21}$$

3.2.8 Analysis of the Duality Gap

Tight Bound on the Relative Duality Gap

The following theorem provides a bound on the relative optimality gap which we can compute in order to assess how far we are from the optimal value.

Theorem 3.2. *Let $f^* > 0$ and $g^* > 0$ given in (3.1) and (3.13) be the optimal values of the primal and dual problems respectively, and let $\hat{f}^* > 0$ given in (3.17) be the computed feasible primal value. Then the relative duality (optimality) gap can be bounded as*

$$0 \leq \frac{g^* - f^*}{f^*} \leq \frac{g^* - \hat{f}^*}{\hat{f}^*} \tag{3.22}$$

Proof. The left inequality follows directly from the positivity of f^* and the weak duality theorem [57, Prop. 5.1.3. p. 495], which states that $g^* \geq f^*$. The right inequality is because $\hat{f}^* \leq f^*$, since $\hat{f}^*$ is a feasible primal value and f^* is the optimal feasible primal value.

We focus on analyzing the absolute duality gap , since the analysis for the relative gap is easily derived by dividing by any feasible solution to the primal problem. Using the optimal λ^* in (3.4e) and (3.17) and substituting it into the numerator of (3.22), we have

$$g^* - \hat{f}^* = \sum_{k \in \mathcal{K}} \mathbb{E}_{\gamma_k} \left\{ w_{m_k^*} \log_2 \left(1 + \tilde{p}_{m_k^*,k}(\lambda^*) \gamma_{m_k^*,k} \right) \right\} + \lambda^* \left(\bar{P} - \hat{P}_{tot}(\lambda^*) \right)$$

$$- \sum_{k \in \mathcal{K}} \mathbb{E}_{\gamma_k} \left\{ w_{m_k^*} \log_2 \left(1 + \tilde{p}_{m_k^*,k}(\lambda^*) \frac{\bar{P}}{\hat{P}_{tot}(\lambda^*)} \gamma_{m_k^*,k} \right) \right\} \tag{3.23}$$

$$\leq \sum_{k \in \mathcal{K}} \mathbb{E}_{\gamma_k} \left\{ w_{m_k^*} \log_2 \left(\frac{1 + \tilde{p}_{m_k^*,k}(\lambda^*) \gamma_{m_k^*,k}}{1 + \tilde{p}_{m_k^*,k}(\lambda^*) \frac{\bar{P}}{\hat{P}_{tot}(\lambda^*)} \gamma_{m_k^*,k}} \right) \right\} + \lambda^* \left(\bar{P} - \hat{P}_{tot}(\lambda^*) \right)$$

where m_k^* in this context is the "winning user" for each possible realization of γ_k. We used the summation across k to encompass the case wherein γ_ks are not identically distributed. Notice that if $P_{tot}(\lambda^*) = \bar{P}$, i.e. if our dual optimal power satisfy the power constraint tightly, the duality gap upper bound is zero, thus the dual optimal and primal optimal solutions are equal and we have solved our problem exactly. This gives us the following corollary:

Corollary 3.3. *If $\exists \lambda^* > 0$ a solution to (3.13) such that $P_{tot}(\lambda^*) = \bar{P}$, then the duality gap is zero, i.e. $f^* = g^*$, and solving the dual problem also solves the primal problem.*

Unfortunately, the existence of a λ^* such that $P_{tot}(\lambda^*) = \bar{P}$ cannot be guaranteed in general, since $\hat{P}_{tot}(\lambda^*)$ is a (possibly) discontinuous function of λ, and the discontinuity may actually happen at $\lambda = \lambda^*$ such that the total power does not meet the constraint tightly, i.e. $\hat{P}_{tot}((\lambda^*)^-) > \bar{P} > \hat{P}_{tot}((\lambda^*)^+)$ (note that $\hat{P}_{tot}(\lambda)$ is a non-increasing function of λ). In fact, it has been shown that the discontinuities actually happen at the most interesting places, i.e. at the possible optimal solutions, and thus cannot be ignored [57]. Fig. 3.5 shows the dual objective $\Theta(\lambda)$ (3.4e) and corresponding feasible primal value in $\hat{f}$ (3.17) as a function of λ for a 2-user 4-subcarrier system with $\bar{P} = 1$, user weights $w = [0.4, 0.6]$. We assume an instantaneous rate allocation in this figure for a particular channel realization (see Sec. 3.2.6). Fig. 3.6 shows the same figure with magnification around λ^*. The discontinuity of the primal value at near λ^* is due to the switching of subcarrier allocations at that point, and the non-differentiability of $\Theta(\lambda)$ at λ^* is due to the non-uniqueness of the solution to the dual problem. This happens when there exists two users m, m' such that their marginal duals for a particular subcarrier k are equal at λ^*, i.e. $g_{m,k}(\gamma_{m,k}, \lambda^*) = g_{m',k}(\gamma_{m',k}, \lambda^*)$; but the resulting power allocation are unequal $\tilde{p}_{m,k} \neq \tilde{p}_{m',k}$. Hence, this causes a jump discontinuity of the sum power around the constraint $\bar{P}$. This phenomenon is illustrated in Fig. 3.7 for the same experiment.

Fortunately, the height of the discontinuity (if it exists) is quite small, and actually diminishes quickly as K increases, and thus the duality gap also diminishes quickly. A heuristic explanation for this phenomenon lies in the fact that as more and more subcarriers are available to sum to $\hat{P}_{tot}(\lambda)$ (c.f. (3.16)), the smaller the contribution of each particular term to the total power, and hence the height of a possible discontinuity likewise becomes smaller. The quantitative behavior of the duality gap bound as K increases has been shown in [43] for the instantaneous continuous rate case with perfect CSI.

General Bound on the Duality Gap

The effect of K and the number of constraints on the duality gap can also be analyzed in a more general framework. An analytical bound for the duality gap of separable integer programming problems has been derived in [70, Prop. 5.26] as

$$g^* - f^* \leq (C+1) \max_{k \in \mathcal{K}} \{\rho_k\} \tag{3.24}$$

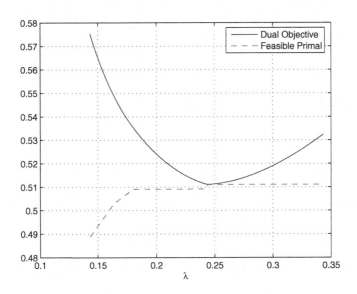

Fig. 3.5. Primal and dual values as a function of λ.

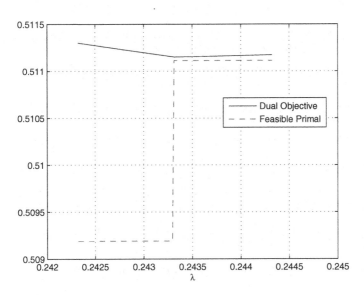

Fig. 3.6. Primal and dual values as a function of λ magnified around λ^*.

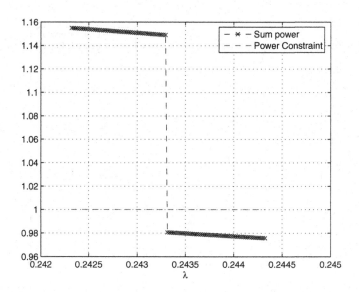

Fig. 3.7. Sum power discontinuity example for a 2-user 4-subcarrier system with $\bar{P} = 1$.

where C is the number of dualized constraints, and ρ_k is a constant for each separable term in the objective that characterizes "how far from convex" our problem is. For our problem, $C = 1$ since we only have a total power constraint; and the constants can be bounded as [70, p. 372]

$$
\begin{aligned}
\rho_k \leq \ &\max_{\boldsymbol{p}_k \in \mathcal{P}_k} \sum_{m \in \mathcal{M}} w_m \mathbb{E}_{\gamma_{m,k}} \left\{ R_{m,k} \left(p_{m,k} \gamma_{m,k} \right) \right\} \\
&- \min_{\boldsymbol{p}_k \in \mathcal{P}_k} \sum_{m \in \mathcal{M}} w_m \mathbb{E}_{\gamma_{m,k}} \left\{ R_{m,k} \left(p_{m,k} \gamma_{m,k} \right) \right\}
\end{aligned}
\tag{3.25}
$$

Since we cannot allocate more than the total power $\bar{P}$ for any subcarrier, the first term in (3.25) can be simplified to $\max_{m \in \mathcal{M}} w_m \mathbb{E}_{\gamma_{m,k}} \left\{ R_{m,k} \left(\bar{P} \gamma_{m,k} \right) \right\}$. Also, we can have zero power allocated to a subcarrier; thus, the second term is zero. Hence, (3.25) simplifies to

$$
\rho_k \leq \max_{m \in \mathcal{M}} w_m \mathbb{E}_{\gamma_{m,k}} \left\{ R_{m,k} \left(\bar{P} \gamma_{m,k} \right) \right\}
\tag{3.26}
$$

Plugging (3.26) back into (3.24), we have

$$
g^* - f^* \leq 2 \max_{m \in \mathcal{M}, k \in \mathcal{K}} w_m \mathbb{E}_{\gamma_{m,k}} \left\{ R_{m,k} \left(\bar{P} \gamma_{m,k} \right) \right\}
\tag{3.27}
$$

which can be interpreted as twice the maximum weighted conditional expected rate over all users and subcarriers when *all* the power is allocated to it.

Although quite loose, the significance of this bound lies in two important observations:

1. *The absolute duality gap bound **does not** scale with K*
   If we include the bandwidth term B/K into the per-subcarrier rate[4], it can be seen that the duality gap diminishes as $K \to \infty$. A similar observation has also been made in [58], using an argument based on the correlation of channel gains for adjacent subcarriers. The diminishing of the bound in (3.24) only relies on the problem structure in multicarrier transmission, which typically has a large number of subcarriers K (e.g. $K > 128$ [17] [19]) and a small number of constraints (see the next item below). Thus, (3.24) generalizes similar observations in [42] [58].

2. *The absolute duality gap bound scales linearly with the number of dualized constraints*
   This fact emphasizes the suitability of this framework to downlink OFDMA and other multiuser multicarrier problems since the number of subcarriers are typically chosen to be much larger than the number of users $K \gg M$, and the number of constraints typically scales with the number of users (e.g. $C = M$ in uplink OFDMA and $C = 2M$ in uplink OFDMA with rate constraints, but $C = 1$ in downlink OFDMA). However, it is more difficult to achieve a certain target duality gap in problems with more dualized constraints, and may thus require more iterations to solve.

Hence, it is the ratio of separable terms to the number of constraints K/C, the ease in which the dual objective is computed, and the existence of good heuristics to map a dual optimal solution to a feasible primal solution that dictate the suitability of the dual optimization framework to a particular problem. Fortunately, multicarrier resource allocation problems often lie in these categories, and are thus prime candidates for using the dual optimization framework.

Finally, by using (3.27) in the relative gap formula given by (3.22), and noticing that

$$f^* \geq \max_{m \in \mathcal{M}, k \in \mathcal{K}} w_m \mathbb{E}_{\gamma_{m,k}} \left\{ R_{m,k} \left(\bar{P} \gamma_{m,k} \right) \right\}$$

since $\max_{m \in \mathcal{M}, k \in \mathcal{K}} w_m \mathbb{E}_{\gamma_{m,k}} \left\{ R_{m,k} \left(\bar{P} \gamma_{m,k} \right) \right\}$ is a feasible solution, we have the following proposition that presents a general, albeit very loose, upper bound on the relative duality gap .

Proposition 3.4. *The relative duality gap for (3.1) is bounded by*

$$0 \leq \frac{g^* - f^*}{f^*} \leq 2 \tag{3.28}$$

---

[4] We excluded this term from the problem formulation for notational brevity, since it is just a constant that does not affect the optimization problem.

3.3 Discrete Rate Maximization with perfect CSI and CDI

3.3.1 Problem Formulation

In this section, we derive resource allocation algorithms for the practically relevant case of when only a discrete number of modulation and coding levels are available (i.e. adaptive modulation and coding).

The average discrete weighted-sum rate maximization can be formulated as

$$f_d^* = \max_{\boldsymbol{p} \in \mathcal{P}} \quad \mathbb{E}_\gamma \left\{ \sum_{m \in \mathcal{M}} w_m \sum_{k \in \mathcal{K}} R_{m,k}^d (p_{m,k} \gamma_{m,k}) \right\}$$

$$\text{s.t.} \quad \mathbb{E}_\gamma \left\{ \sum_{m \in \mathcal{M}} \sum_{k \in \mathcal{K}} p_{m,k} \right\} \leq \bar{P} \tag{3.29}$$

where $R_{m,k}^d (p_{m,k} \gamma_{m,k})$ is the discrete rate function given in (2.8).

3.3.2 Dual Optimization Framework

Following a dual optimization framework that is similar to Section 3.2.2, we arrive at the dual objective (c.f. (3.4e))

$$\Theta_d(\lambda) = \lambda \bar{P} + K \mathbb{E}_\gamma \left\{ \max_{m \in \mathcal{M}} \left[\max_{p_{m,k} \geq 0} \left(w_m R_{m,k}^d (p_{m,k} \gamma_{m,k}) - \lambda p_{m,k} \right) \right] \right\} \tag{3.30}$$

The main difference of the inner maximization in this case with the continuous rate case in (3.4e) is that $R_{m,k}^d (p_{m,k} \gamma_{m,k})$ is a discontinuous function; hence, simple differentiation to arrive at the optimal solution is not feasible. However, note that we can divide the feasible region for $p_{m,k}$ (i.e. the non-negative real line) into L segments $\mathcal{R}_+^l = \left[\frac{\eta_l}{\gamma_{m,k}}, \frac{\eta_{l+1}}{\gamma_{m,k}} \right), l \in \mathcal{L} \equiv \{0, \dots, L-1\}$. Since λ and $p_{m,k}$ are both non-negative, we have

$$w_m R_{m,k}^d (p_{m,k} \gamma_{m,k}) - \lambda p_{m,k} = w_m r_l - \lambda p_{m,k}$$
$$\leq w_m r_l - \lambda \frac{\eta_l}{\gamma_{m,k}}, \quad \forall p_{m,k} \in \mathcal{R}_+^l \tag{3.31}$$

Thus, there are only L candidate power allocation functions

$$\tilde{p}_{m,k}^d \in \left\{ \frac{\eta_0}{\gamma_{m,k}}, \dots, \frac{\eta_{L-1}}{\gamma_{m,k}} \right\} \tag{3.32}$$

from which we need to choose the one that maximizes $w_m r_l - \lambda \frac{\eta_l}{\gamma_{m,k}}$, i.e.

$$\tilde{p}_{m,k}^d = \frac{\eta_{l_{m,k}^*}}{\gamma_{m,k}} \tag{3.33}$$

where

$$l^*_{m,k} \in \arg \max_{l \in \mathcal{L}} \left(w_m r_l - \lambda \frac{\eta_l}{\gamma_{m,k}} \right) \tag{3.34}$$

We call (3.33) a *multi-level fading inversion* (MFI) power allocation, since it is simply the inverse of the fading CNR scaled by the different SNR transitions η_l. This in turn also gives us the rate allocation $\tilde{R}^d_{m,k} = r_{l^*_{m,k}}$.

A straightforward computation of (3.34) would require $\mathcal{O}(L)$ complexity. However, if we assume that the discrete rate function $R^d(p_{m,k}\gamma_{m,k})$ is concave[5], we can reduce the complexity of finding the power allocation function by noticing that (3.34) is equivalent to (see Appendix C for a derivation)

$$l^*_{m,k} = \left\{ l \in \mathcal{L} : \frac{\lambda}{w_m \gamma_{m,k}} \in \left[\frac{r_{l+1} - r_l}{\eta_{l+1} - \eta_l}, \frac{r_l - r_{l-1}}{\eta_l - \eta_{l-1}} \right) \right\} \tag{3.35}$$

where with slight abuse of notation, we define $(r_0 - r_{-1})/(\eta_0 - \eta_{-1}) \equiv \infty$. This can be interpreted geometrically by treating $\frac{\lambda}{w_m \gamma_{m,k}}$ as a slope value for which we are looking for an interval of consecutive slope values for which it belongs (see [33] for a similar interpretation for single-user discrete multitone systems). Since the set of rates and SNR region boundaries r_l and η_l are predefined in a communications system, we can store the set of slopes into a lookup table, thereby reducing the complexity of finding the optimal power to a single table lookup operation. Fig. 3.8 shows an example of the slope searching procedure for the discrete rate function given in Fig. 2.1. In this example, $\lambda/(w_m \gamma_{m,k})$ is the slope of the dashed line, and has a value that is between the minimum and maximum slope values for rate level $l = 2$. Therefore, $l^*_{m,k} = 2$, and the rate allocation we choose is $r^*_{m,k} = 4$ which is 16-QAM.

Finally, we can write the discrete rate maximization dual problem as

$$g^*_d = \min_{\lambda \geq 0} \lambda \bar{P} + K \mathbb{E}_{\gamma_k} \left\{ g^d_k(\gamma_k, \lambda) \right\} \tag{3.36}$$

$$g^d_k(\gamma_k, \lambda) = \max_{m \in \mathcal{M}} \left\{ g^d_{m,k}(\gamma_{m,k}, \lambda) \right\} \tag{3.37}$$

where (3.37) is a max function over the M per-subcarrier dual functions given as

$$g^d_{m,k}(\gamma_{m,k}, \lambda) = \max_{l \in \mathcal{L}} \left\{ w_m r_l - \lambda \frac{\eta_l}{\gamma_{m,k}} \right\} \tag{3.38}$$

Note that despite the negative term in (3.38), $g^d_{m,k}(\gamma_{m,k}, \lambda)$ is always non-negative. This is because both r_0 and η_0 are equal to zero; hence, the lowest possible value for the objective is zero.

---

[5] Concavity for this discontinuous staircase function simply means that the slopes when "connecting the dots" of the edges of the staircase are non-increasing.

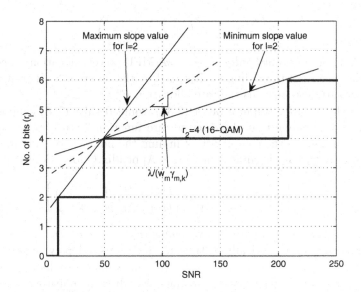

Fig. 3.8. Example of the slope searching procedure where $l_{m,k}^* = 2$.

3.3.3 Numerical Evaluation of the Expected Dual

Similar to the continuous rate case (cf. 3.2.3), we require an M-dimensional integral to compute the expectation in (3.36) in a straightforward manner. Thus, we proceed similarly as the continuous rate case to derive a closed-form expression for the pdf of g_k^d in (3.37) and reduce the computation to just a single integral. The key to the derivation is to derive the CDF and PDF of (3.38), and use the same formula used in the continuous rate case for the maximum order statistic given in (3.9). Making the same assumption that the discrete rate function $R_{m,k}^d$ (2.8) is concave, the CDF and PDF are given as (see Appendix D for a derivation)

$$F_{g_{m,k}^d}(g_{m,k}^d) = u(g_{m,k}^d)F_{\gamma_{m,k}}(s_1) + \sum_{l \in \mathcal{L}\backslash 0} \left[F_{\gamma_{m,k}}\left(\min\left(h_l(g_{m,k}^d), s_{l+1}\right)\right) - F_{\gamma_{m,k}}(s_l) \right]^+$$

(3.39)

$$f_{g_{m,k}^d}(g_{m,k}^d) = \delta(g_{m,k}^d)F_{\gamma_{m,k}}(s_1) +$$
$$\sum_{l \in \mathcal{L}\backslash 0} \mathbf{1}\left(s_l \leq h_l(g_{m,k}^d) \leq s_{l+1}\right) f_{\gamma_{m,k}}\left(h_l(g_{m,k}^d)\right) \frac{h_l^2(g_{m,k}^d)}{\lambda \eta_l}$$

(3.40)

where $h_l(g_{m,k}^d) = \frac{\lambda \eta_l}{[w_m r_l - g_{m,k}^d]^+}$, $s_l = \frac{\lambda(\eta_l - \eta_{l-1})}{w_m(r_l - r_{l-1})}$. Figs. 3.9-3.10 shows an example of the cdf and pdf for $w_m = 1$, $\lambda = 1$, $\bar{\gamma} = 20$ dB, and discrete rate function given in Fig. 2.1. We also plot the L individual terms that sum

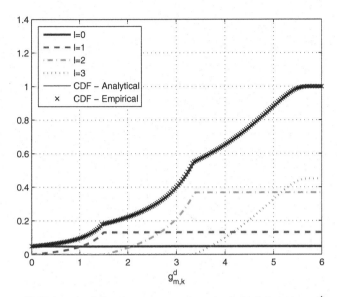

Fig. 3.9. CDF (3.39) of the discrete rate marginal dual function $g_{m,k}^d$ (3.38).

to the functions, thereby giving better insight into how these functions are derived. We also superimposed empirical curves generated using Monte-Carlo generation for verification.

Finally, by using (3.39)-(3.40) in (3.9) and then in (3.36), the dual problem can be written as

$$g_d^* = \min_{\lambda \geq 0} \left[\lambda \bar{P} + K \int_0^{\infty} g_k^d f_{g_k^d}(g_k^d) dg_k^d \right] \qquad (3.41)$$

3.3.4 Optimal Discrete Rate, Subcarrier, and Power Allocation

The optimum solution to (3.41) denoted by λ^* can be found using similar line search techniques. The optimal subcarrier, rate, and power allocation is then determined using λ^* as

$$m_k^* = \arg\max_{m \in \mathcal{M}} w_m r_{l_{m,k}^*} - \lambda^* \frac{\eta_{l_{m,k}^*}}{\gamma_{m,k}} \qquad (3.42)$$

$$R_{m,k}^* = r_{l_{m,k}^*} \mathbf{1}(m = m_k^*) \qquad (3.43)$$

$$p_{m,k}^* = \frac{\eta_{l_{m,k}^*}}{\gamma_{m,k}} \mathbf{1}(m = m_k^*) \qquad (3.44)$$

where $l_{m,k}^*$ is given by (3.35) with $\lambda = \lambda^*$. An upper bound on the relative duality gap of this algorithm can be derived similarly to Section 3.2.8. The

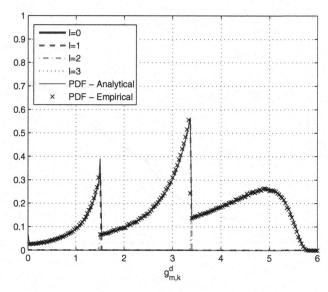

Fig. 3.10. PDF (3.40) of the discrete rate marginal dual function $g^d_{m,k}$ (3.38).

complexity analysis is also similar to Sec. 3.2.5, except for the additional $\mathcal{O}(L)$ factor to compute the dual objective in (3.41), giving an initialization complexity of $\mathcal{O}(INML)$; and the additional $\mathcal{O}(\log(L))$ for the table lookup operation in (3.35), thereby giving a resource allocation complexity of $\mathcal{O}(MK\log(L))$. Fig. 3.11 presents a flow chart of the OFDMA ergodic weighted-sum discrete rate maximization algorithm. See Table 3.4 for a comparison between continuous and discrete rate resource allocation algorithms in terms of initialization and per-symbol complexity.

The instantaneous discrete rate maximization algorithm can also be derived by solving for the optimal instantaneous geometric multiplier λ^*_{inst} using (3.36) without the expectation and using the actual CNR vector $\boldsymbol{\gamma}$. The allocation rules are also given by (3.14)-(3.15) using the multiplier λ^*_{inst}. A further simplification is to assume constant power allocation, where the user selection is

$$m^*_k = \arg \max_{m \in \mathcal{M}} w_m R^d_{m,k} \left(\frac{\bar{P}}{K} \gamma_{m,k} \right) \qquad (3.45)$$

3.4 Numerical Results

We consider an OFDMA system based on a 3GPP-LTE downlink [19] with parameters given in Table 3.1. We simulate the frequency-selective Rayleigh fading channel using the ITU-Vehicular A channel model [71]. For each user's channel realization $\boldsymbol{h}_m$ in (2.3), we generate a complex Gaussian random

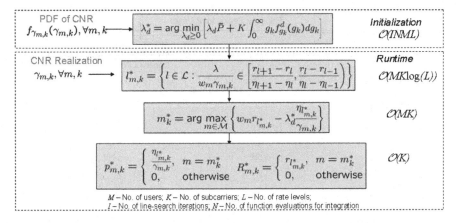

Fig. 3.11. OFDMA resource allocation algorithm for ergodic weighted-sum discrete rate maximization.

vector with N_t independent entries and variance according to the power delay profile.

Table 3.1. Simulation Parameters

Parameter	Value
Subcarriers (K_{fft})	128
Used Subcarriers (K)	76
Bandwidth (B)	1.25 MHz
Sampling Freq. (F_s)	1.92 MHz
Cylic Prefix Length (L_{cp})	6 samples
Average Power Constraint ($\bar{P}$)	1

3.4.1 Continuous Rate Allocation

In Fig. 3.12, we compare the capacity regions for the continuous rate allocation case with 2 users using $10,000$ channel realizations and varying w_1 between 0 and 1, and setting $w_2 = 1 - w_1$. We see that ergodic rates maximization has better performance than the instantaneous rate and constant power allocation cases due to its ability to exploit the temporal dimension[6]. The gain is also more pronounced for lower SNRs and more disparate user weights, which is analogous to previous studies in adaptive modulation, e.g. [38] [24], which

---

[6] This is accomplished through knowledge of the fading distribution, and flexibility in allocating the total power per symbol as long as the *average* power constraint is fulfilled.

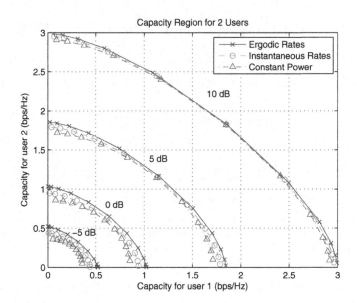

Fig. 3.12. Two-user capacity region for ergodic and instantaneous continuous rate maximization and constant power allocation.

concluded that the exploitation of the additional temporal dimension through the ergodic formulation is most useful when other degrees of freedom have been significantly curtailed.

In Fig. 3.13, we show the resulting average power allocation for ergodic rate maximization (computed via numerical integration) to each of the two users as the weights are being varied. We can see that the sum of both user powers is equal to unity almost exactly, thus satisfying the power constraint tightly. Thus, we expect the relative duality gap to be almost zero. In this case, our average relative duality gap is 2.0491×10^{-6}.

Fig. 3.14 shows the sum capacity as the number of users, M, is increased. We ran 500 frames with 1000 symbols per frame, where we draw a random realization of the normalized user weights w_m and hold it constant for each frame. We see the effect of *multiuser diversity* in that the capacity is actually increasing as the number of users increase. The gain of ergodic rates over the other methods diminish as we increase M, which is consistent with [38].

In Table 3.2, we present other relevant metrics for the continuous rate maximization algorithms. For the ergodic rate maximization, the first main column indicates the average number of function evaluations required to numerically compute the integration of (3.13) with a tolerance of 10^{-6}, and the second main column indicates the average number of Golden-section search iterations to solve for λ^* in the dual problem (3.6) with a tolerance of 10^{-4}. Note that this computation is performed only once during initialization and

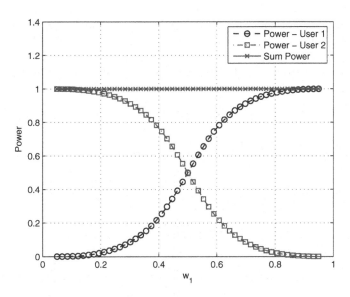

Fig. 3.13. Average power allocated to the two users in ergodic rate maximization.

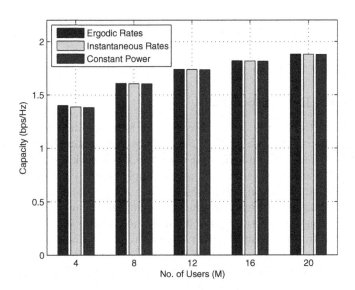

Fig. 3.14. Sum capacity for different numbers of users.

does not need to be performed while the pdf of the channel fading remains constant. The second column for instantaneous rate maximization is the average number of iterations for each channel realization. The third column for both cases is the relative duality gap upper bound computed by (3.22). Note that the duality gaps are negligible, and thus both algorithms can be considered optimal. Since the constant power allocation does not involve iterations, it is not included in Table 3.2.

Table 3.2. Relevant Performance Measures for the Continuous Rate Resource Allocation Algorithms

Measure	SNR	Ergodic	Instantaneous
[a]No. of Fun. Eval. (N)	5 dB	47.912	—
	10 dB	50.091	—
	15 dB	53.732	—
[b]No. of Iterations (I)	5 dB	8.091	8.344
	10 dB	7.727	8.333
	15 dB	7.936	8.539
[c]Relative Gap ($\times 10^{-6}$)	5 dB	7.936	0.025
	10 dB	5.462	0.023
	15 dB	5.444	0.016

[a] Average no. of function evaluations for numerical integration in (3.13)
[b] No. of line search iterations to solve the dual problem in (3.13) or (3.19)
[c] Relative duality/optimality gap given in (3.22)

3.4.2 Discrete Rate Allocation

Fig. 3.15 shows the results of the discrete rate resource allocation using the discrete rate function given in Sec. 2.4.3. Note that channel coding is not present in this case for simplicity, but since the framework merely needs the SNR thresholds and rate values, the results apply to the coded case as long as the discrete rate function is concave.

Note that the general trends are similar to the continuous rate case, except that the advantage for ergodic rates is much more pronounced, and a large loss is incurred by the constant power allocation case. This is due to the big loss of freedom in the rate allocation (limited to just 4 rates in contrast to an infinite number of rates in the continuous rate case), which when coupled with constant power allocation results in a huge loss in performance.

Fig. 3.16 shows the sum rates as the number of users is increased for the three different methods using a similar simulation setup as in the continuous rates case. We see similar trends as in the continuous rates case, but also with more pronounced gains for the ergodic rates case.

Table 3.3 shows the average number of iterations and the relative optimality gaps for the discrete rate allocation algorithms. Note that the number of

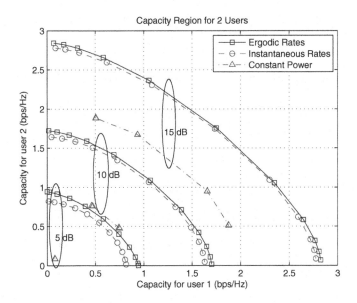

Fig. 3.15. Two-User capacity region for ergodic and instantaneous discrete rate maximization.

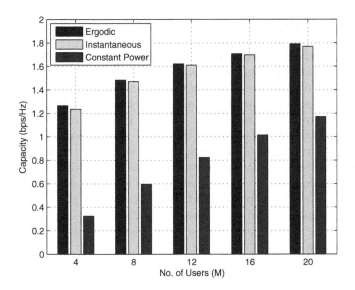

Fig. 3.16. Sum rates for different numbers of users.

function evaluations are higher, due primarily to the discontinuities in the cdf and pdf functions (see Figs. 3.9-3.10).

Table 3.3. Relevant Performance Measures for the Discrete Rate Resource Allocation Algorithms

Measure	SNR	Ergodic	Instantaneous
[a]No. of Fun. Eval. (N)	5 dB	62.09	−
	10 dB	91.55	−
	15 dB	133.02	−
[b]No. of Iterations (I)	5 dB	9.818	17.241
	10 dB	10.550	17.200
	15 dB	9.909	17.304
[c]Relative Gap ($\times 10^{-4}$)	5 dB	0.871	3.602
	10 dB	0.951	1.038
	15 dB	0.532	0.340

[a] Average no. of function evaluations for numerical integration in (3.41)
[b] Average no. of line search iterations to solve the dual problem in (3.41)
[c] Average relative duality/optimality gap given in (3.22) but for the discrete rate case

3.4.3 Complexity Comparison

Table 3.4 shows the complexity order of the different resource allocation algorithms[7]. If we use the average numbers given in Tables 3.2 and 3.3, the ergodic rate algorithms are less complex than the instantaneous rate algorithms per symbol on average, as long as the rate of change of the channel fading statistics (roughly at the rate of change of slow fading, e.g. Log-normal shadowing) is much lower than the rate of change of the actual channel realizations (roughly at the rate of fast fading, e.g. Rayleigh fading), such that the initialization is performed less often. One caveat, however, is that the ergodic rate algorithms require information on the channel fading distribution function, which need an additional level of complexity and feedback overhead. Furthermore, the peak-to-average power ratio of the power allocation in the ergodic rates case is typically higher than for instantaneous rates, and even more so for constant power allocation.

---

[7] Note that the complexity analyzed here is purely from the resource allocation perspective, and does not include actual transmission and decoding complexity. In order to achieve the ergodic (Shannon) capacity (continuous rates case), random coding with infinite block lengths are required [28], and is impractical from an implementation perspective.

Table 3.4. Comparison of the proposed ergodic and instantaneous rate resource allocation algorithms with constant power allocation algorithm assuming perfect CSI and CDI.

Algorithm	Initialization Complexity	Per-symbol Complexity	Data Rates	Rel. Gap Order
Cont. Rates, Ergodic	$\mathcal{O}(INM)$	$\mathcal{O}(MK)$	High	10^{-6}
Cont. Rates, Inst.	–	$\mathcal{O}(IMK)$	High	10^{-8}
Cont. Rates, Const. Pow.	–	$\mathcal{O}(MK)$	High	–
Disc. Rates, Ergodic	$\mathcal{O}(INML)$	$\mathcal{O}(MK\log(L))$	Med.	10^{-5}
Disc. Rates, Inst.	–	$\mathcal{O}(IMK\log(L))$	Med.	10^{-4}
Disc. Rates, Const. Pow.	–	$\mathcal{O}(MK\log(L))$	Low	–

M-no. of users, K-no. of subcarriers, L-No. of discrete rates,
N-no. of function evaluations for integration, I-no. of line search iterations.

3.5 Conclusion

In this chapter, we derived the optimal resource allocation algorithms for continuous and discrete ergodic weighted-sum rate maximization in OFDMA systems assuming perfect CSI and CDI. The algorithms are based on a dual optimization framework with per-symbol complexity of $\mathcal{O}(MK)$ per iteration, and are shown to achieve relative optimality gaps of less than 10^{-4} using 3GPP-LTE OFDMA simulation parameters. It is also shown that ergodic rate maximization is actually less complex per symbol than instantaneous rate maximization, and thus presents an attractive communication performance vs. complexity tradeoff. The most gain in ergodic maximization occur at low SNRs and for discrete rate cases, primarily because of decreased degrees of freedom in these scenarios.

The following are the main contributions of this chapter:

- *Optimal resource allocation in continuous rate case:* Established that the optimal subcarrier and power allocation in ergodic and instantaneous weighted-sum continuous rate maximization is *multi-level waterfilling* with *max-dual user selection,* and the resource allocation procedure is parameterized by a single geometric multiplier.
- *Optimal resource allocation in discrete rate case:* Established that the optimal subcarrier, rate, and power allocation in ergodic and instantaneous weighted-sum discrete rate maximization is *multi-level fading inversion* with *max-dual rate and user selection,* and the resource allocation procedure is likewise parameterized by a single geometric multiplier.
- *Linear complexity algorithms for resource allocation:*Derived efficient linear complexity algorithms for finding the optimal geometric multipliers for both continuous and discrete rates which entail a simple line search and a single integral for each function evaluation of the line search.
- *Duality gap analysis:* Analyzed the duality gap for the continuous rates case, and established easily verifiable conditions for the existence or non-existence of a duality gap.

4

Weighted-Sum Rate Maximization with Partial CSI

4.1 Introduction

In the previous chapter, we studied OFDMA resource allocation algorithms assuming perfect CSI and CDI. This is actually a common underlying assumption among most previous research in OFDMA resource allocation (see Sec. 2.2). This assumption is quite unrealistic due to channel estimation errors, and more importantly, channel feedback delay. In this chapter, we focus on the weighted-sum rate maximization where only imperfect (partial) CSI is available, but where the CDI of the partial CSI is still known. The contents of this chapter are close to that of the papers [72] [73] [74].

The effect of imperfect CSI for rate maximization in wireless systems has been quite well studied for single-user wireless systems. In [75], adaptive trellis-coded modulation schemes using a single outdated channel estimate for single-carrier systems in flat-fading channels were proposed. In [76], uncoded adaptive modulation schemes using predicted CSI were developed, also for single-carrier flat-fading channels. In [77] [78], the effect of channel estimation errors and channel feedback delay on adaptive modulation for OFDM systems in time and frequency selective channels was studied. It was concluded that the detrimental effect of outdated channel information is significant, and that using OFDM channel prediction [77] [79] [80] [81] [82] [83] or using multiple channel estimates [78] is a viable way of overcoming this delay. In [84], power allocation methods for ergodic and outage capacity maximization in OFDM were studied assuming that the partial CSI distribution information is available. Adaptive modulation in single-user single-carrier Multiple-Input, Multiple-Output (MIMO) systems [85] [86] and MIMO-OFDM systems [87] assuming imperfect or predicted CSI have also been investigated. However, no work to the best of the authors's knowledge has considered the multiuser OFDM case.

In the previous chapter it was shown that by using a dual optimization approach, the perfect CSI problem can be solved with just $\mathcal{O}(MK)$ complexity per symbol for an OFDMA system with M active users and K used subcarriers. Using a similar dual optimization approach, we relax the assumption of perfect CSI in this chapter, and formulate and solve the problem assuming the availability of imperfect CSI. We use the statistics of this imperfect CSI to perform resource allocation for both continuous rate (capacity based) and discrete rate (adaptive modulation and coding based) maximization cases. We considered minimum mean square error (MMSE) OFDM channel prediction in this chapter, but the framework can be easily extended to other estimation/prediction approaches as well. We show that by using the dual optimization framework, we can solve the imperfect CSI problem with relative optimality gaps of less than 10^{-5} for continuous rates and less than 10^{-3} for discrete rates in cases of practical interest.

This chapter is organized as follows. Section 4.2 discusses the partial CSI model used in this chapter. Section 4.3 discusses the optimal resource allocation algorithms for the continuous rate case (ergodic (Shannon) capacity) assuming partial CSI. Section 4.4 considers the more practically relevant case of allocation for discrete rates (adaptive modulation and coding). Section 4.5 presents several numerical examples based on a 3GPP-LTE downlink OFDMA system, and Section 4.6 concludes this chapter.

4.2 Partial Channel State Information Model

Suppose we wish to perform resource allocation for the mth user with actual fading channel vector $\boldsymbol{h}_m$ at symbol index n, but only P symbols of delayed and noisy estimates of the channel D_t apart are available, which we denote as

$$\widetilde{\boldsymbol{h}}_m[n - pD_t] = \boldsymbol{h}_m[n - pD_t] + \boldsymbol{e}_m[n - pD_t], \quad p = 1, \dots, P \qquad (4.1)$$

where $\boldsymbol{e}_m[n - pD_t] \sim \mathcal{CN}(\boldsymbol{0}_K, \sigma_e^2 \mathbf{I}_K)$ is the spectrally and temporally white estimation error random vector with estimation error variance σ_e^2 which is uncorrelated with $\boldsymbol{h}_m[n - pD_t]$. This can effectively model a least-squares estimate of the channel using pilot tones with power σ_t^2, resulting in $\sigma_e^2 = \sigma_\nu^2 / \sigma_t^2$. Stacking these into a PK-length vector, which we denote as

$$\widetilde{\mathfrak{h}}_m = \left[\widetilde{\boldsymbol{h}}_m^T[n - D_t], \widetilde{\boldsymbol{h}}_m^T[n - 2D_t], \dots, \widetilde{\boldsymbol{h}}_m^T[n - PD_t] \right]^T,$$

results in a ZMCSCG random vector with $PK \times PK$ block Hermitian-Toeplitz covariance matrix

$$\boldsymbol{\Sigma}_{\widetilde{\mathfrak{h}}_m} = \begin{bmatrix} \mathbf{W}\mathbf{R}_m[0]\mathbf{W}^H + \sigma_e^2\mathbf{I} & \mathbf{W}^H\mathbf{R}_m^H[D_t]\mathbf{W} & \cdots & \mathbf{W}^H\mathbf{R}_m^H[(P-1)D_t]\mathbf{W} \\ \mathbf{W}\mathbf{R}_m[D_t]\mathbf{W}^H & \mathbf{W}\mathbf{R}_m[0]\mathbf{W}^H + \sigma_e^2\mathbf{I} & \cdots & \mathbf{W}^H\mathbf{R}_m^H[(P-2)D_t]\mathbf{W} \\ \vdots & \vdots & \ddots & \vdots \\ \mathbf{W}\mathbf{R}_m[(P-1)D_t]\mathbf{W}^H & \mathbf{W}\mathbf{R}_m[(P-2)D_t]\mathbf{W}^H & \cdots & \mathbf{W}\mathbf{R}_m[0]\mathbf{W}^H + \sigma_e^2\mathbf{I} \end{bmatrix}$$

$$(4.2)$$

where $\mathbf{R}_m$ is the Hermitian-symmetric and Toeplitz $P \times P$ temporal autocorrelation matrix with entries $[\mathbf{R}_m]_{i,j} = r_m[(i-j)D_t]$. The conditional distribution of the desired channel is then

$$h_m | \tilde{\mathfrak{h}}_m \sim \mathcal{CN}\left(\hat{h}_m, \hat{\Sigma}_m\right) \tag{4.3}$$

where

$$\hat{h}_m = \Sigma_{h_m \tilde{\mathfrak{h}}_m} \Sigma_{\tilde{\mathfrak{h}}_m}^{-1} \tilde{\mathfrak{h}}_m \tag{4.4}$$

is the conditional mean estimator, which is also the MMSE predictor for the channel [69];

$$\hat{\Sigma}_m = \Sigma_{h_m} - \Sigma_{h_m \tilde{\mathfrak{h}}_m} \Sigma_{\tilde{\mathfrak{h}}_m}^{-1} \Sigma_{h_m \tilde{\mathfrak{h}}_m}^{H} \tag{4.5}$$

is the conditional covariance, and is also the covariance matrix for the ZMC-SCG prediction error vector we denote as $\hat{e}_m$, i.e.

$$\hat{e} \sim \mathcal{CN}(\mathbf{0}_K, \hat{\Sigma}_m) \tag{4.6}$$

and

$$\Sigma_{h_m \tilde{\mathfrak{h}}_m} = \mathbf{W}\left[\mathbf{R}_m[D_t], \ldots, \mathbf{R}_m[PD_t]\right] \mathbf{W}^H \tag{4.7}$$

is the $K \times PK$ cross-covariance matrix. Interestingly, the probability density function (pdf) of h_m given the MMSE estimate $\hat{h}_m$ is identical to $h_m | \tilde{\mathfrak{h}}_m$, i.e.

$$h_m | \hat{h}_m \sim \mathcal{CN}\left(\hat{h}_m, \hat{\Sigma}_m\right) \tag{4.8}$$

If we assume identical normalized temporal autocorrelation functions across multipath taps, then we can simplify (4.5) to

$$\hat{\Sigma}_m = \Sigma_{h_m} - \left(\mathbf{r}_m^T \otimes \Sigma_{h_m}\right) \left(\mathbf{R}_m \otimes \Sigma_{h_m} + \sigma_e^2 \mathbf{I}_{PK}\right)^{-1} \left(\mathbf{r}_m^T \otimes \Sigma_{h_m}\right)^H \tag{4.9}$$

where $\mathbf{r}_m^T = [r_m[D_t], \ldots, r_m[PD_t]]$ and $\otimes$ is the Kronecker product.

Note that $\hat{h}_m$ is also ZMCSCG with covariance $\Sigma_{\hat{h}_m} = \Sigma_{h_m} - \hat{\Sigma}_m$, i.e.

$$\hat{h}_m \sim \mathcal{CN}\left(\mathbf{0}_K, \Sigma_{h_m} - \hat{\Sigma}_m\right) \tag{4.10}$$

Therefore, assuming MMSE channel prediction, we can write the perfect CSI vector as the sum of the predicted channel $\hat{h}_m$ and the prediction error $\hat{e}_m$ which are uncorrelated with each other

$$h_m = \hat{h}_m + \hat{e}_m \tag{4.11}$$

We shall use this equation to generate both the partial CSI and perfect CSI in the results section (Section 4.5).

In ergodic capacity maximization with imperfect CSI, we require the marginal distribution for each subcarrier. The marginal fading distribution on subcarrier k conditioned on the estimated channels is a non-zero mean complex Gaussian random variable given as $h_{m,k}|\hat{h}_{m,k} \sim \mathcal{CN}(\hat{h}_{m,k}, \hat{\sigma}^2_{m,k})$ where $\hat{h}_{m,k}$ is the kth element in $\hat{\boldsymbol{h}}_m$ and $\hat{\sigma}^2_{m,k}$ is the kth diagonal element in $\hat{\boldsymbol{\Sigma}}_m$, which is essentially the prediction error variance for that subcarrier. Thus, the channel-to-noise ratio (CNR) $\gamma_{m,k} = |h_{m,k}|^2/\sigma^2_\nu$ conditioned on $\hat{\gamma}_{m,k} = |\hat{h}_{m,k}|^2/\sigma^2_\nu$ is in turn a non-central Chi-squared (NCχ^2) distributed random variable with two degrees of freedom with pdf

$$f_{\gamma_{m,k}}(\gamma_{m,k}|\hat{\gamma}_{m,k}) = \frac{1}{\rho_{m,k}}e^{-\frac{\hat{\gamma}_{m,k}+\gamma_{m,k}}{\rho_{m,k}}} I_0\left(\frac{2}{\rho_{m,k}}\sqrt{\hat{\gamma}_{m,k}\gamma_{m,k}}\right) \tag{4.12}$$

where I_0 is the zeroth-order modified BesselBessel function of the first kind, and $\rho_{m,k} = \hat{\sigma}^2_{m,k}/\sigma^2_\nu$ is the ratio of the prediction error variance to the ambient noise variance [88, Eq. 2-1-118].

4.3 Continuous Rate Maximization with Partial CSI and CDI

4.3.1 Problem Formulation

We assume that we have knowledge of the imperfect CNR vector $\hat{\boldsymbol{\gamma}} = [\hat{\boldsymbol{\gamma}}_1^T, \ldots, \hat{\boldsymbol{\gamma}}_K^T]^T$, $\hat{\boldsymbol{\gamma}}_k = [\hat{\gamma}_{1,k}, \ldots, \hat{\gamma}_{M,k}]^T$; corresponding to an estimate of the actual CNR realization $\boldsymbol{\gamma} = [\boldsymbol{\gamma}_1^T, \ldots, \boldsymbol{\gamma}_K^T]^T$, where $\boldsymbol{\gamma}_k = [\gamma_{1,k}, \ldots, \gamma_{M,k}]^T$. Further assuming that the conditional distribution of $\gamma_{m,k}|\hat{\gamma}_{m,k}$ is known, the weighted sum rate maximization problem for downlink OFDMA assuming partial CSI is then

$$f^* = \max_{\boldsymbol{p}\in\mathcal{P}} \sum_{m\in\mathcal{M}} w_m \sum_{k\in\mathcal{K}} \mathbb{E}_{\gamma_{m,k}}\left\{R_{m,k}\left(p_{m,k}\gamma_{m,k}\right)|\,\hat{\gamma}_{m,k}\right\}$$
$$\text{s.t.} \quad \sum_{m\in\mathcal{M}}\sum_{k\in\mathcal{K}} p_{m,k} \leq \bar{P} \tag{4.13}$$

where $R_{m,k}$ is given in (2.7), and w_m are the user weights.

Problem Classification

The problem given by (4.13), which is similar to the perfect CSI case in (3.1), is a stochastic mixed-integer programming problem. However, the main difference in this case is that we have replaced the expectation with a conditional expectation given the partial CSI. The optimization variable $\boldsymbol{p}$ in this case is a function of the estimated CNR $\hat{\boldsymbol{\gamma}}$. The main difference in this problem with that of the perfect CSI case in (3.1) is that we no longer need a parametric analysis of the optimal solution as a function of the observation $\hat{\boldsymbol{\gamma}}$ (which is an infinite-dimensional problem, see Sec. 3.2.1), since the average power constraint reduces to a deterministic constraint when given $\hat{\boldsymbol{\gamma}}$. Thus, we need to solve (4.13) for each realization of $\hat{\boldsymbol{\gamma}}$ by searching for $\boldsymbol{p}\in\mathcal{P}$, where $\mathcal{P}$ is defined by (2.6), which is a finite-dimensional problem.

4.3.2 Dual Optimization Framework

The dual problem for (4.13) is defined as

$$g^* = \min_{\lambda \geq 0} \Theta(\lambda) \tag{4.14}$$

$$\Theta(\lambda) = \max_{p \in \mathcal{P}} \sum_{m \in \mathcal{M}} w_m \sum_{k \in \mathcal{K}} \mathbb{E}_{\gamma_{m,k}} \left\{ R_{m,k} \left(p_{m,k} \gamma_{m,k} \right) | \hat{\gamma}_{m,k} \right\} + \lambda \left(\bar{P} - \sum_{k \in \mathcal{K}} \sum_{m \in \mathcal{M}} p_{m,k} \right)$$

$$= \lambda \bar{P} + \sum_{k \in \mathcal{K}} \max_{p_k \in \mathcal{P}_k} \sum_{m \in \mathcal{M}} \left(w_m \mathbb{E}_{\gamma_{m,k}} \left\{ R_{m,k} \left(p_{m,k} \gamma_{m,k} \right) | \hat{\gamma}_{m,k} \right\} - \lambda p_{m,k} \right) \tag{4.15a}$$

$$= \lambda \bar{P} + \sum_{k \in \mathcal{K}} \max_{m \in \mathcal{M}} \max_{p_{m,k} \geq 0} \left(w_m \mathbb{E}_{\gamma_{m,k}} \left\{ R_{m,k} \left(p_{m,k} \gamma_{m,k} \right) | \hat{\gamma}_{m,k} \right\} - \lambda p_{m,k} \right) \tag{4.15b}$$

where (4.15a) follows from the separability of the variables across subcarriers, and (4.15b) from the exclusive subcarrier assignment constraint. The main difference of (4.15b) with the perfect CSI dual in (3.4e) is that we can no longer assume that the marginal CNR conditional distribution across different subcarriers k is identical, since both the estimated CNR and the error variance may be different across subcarriers for a single user.

We denote the optimal power allocation function for the innermost per-user and per-subcarrier problem in (4.15b) as $\tilde{p}_{m,k}(\lambda)$, which can be found using the necessary condition for a constrained optimal solution, and is given as

$$\tilde{p}_{m,k}(\lambda) = \begin{cases} p_{m,k} : \mathbb{E}_{\gamma_{m,k}} \left\{ \frac{\gamma_{m,k}}{1 + p_{m,k} \gamma_{m,k}} \bigg| \hat{\gamma}_{m,k} \right\} = \gamma_{0,m} , & \mathbb{E}_{\gamma_{m,k}} \{ \gamma_{m,k} | \hat{\gamma}_{m,k} \} \geq \gamma_{0,m} \\ 0 & , & \mathbb{E}_{\gamma_{m,k}} \{ \gamma_{m,k} | \hat{\gamma}_{m,k} \} < \gamma_{0,m} \end{cases}$$
$$\tag{4.16}$$

where $\gamma_{0,m} = \frac{\lambda \ln 2}{w_m}$. This can be interpreted as a *multi-level water-filling* with cut-off CNR $\gamma_{0,m}$ similar to the perfect CSI case given in (3.5), except that the cut-off is now based on the conditional mean of the CNR given its estimate, instead of the actual CNR. Using the pdf in (4.12), the conditional mean is simply $\mathbb{E}_{\gamma_{m,k}} \{ \gamma_{m,k} | \hat{\gamma}_{m,k} \} = \hat{\gamma}_{m,k} + \rho_{m,k}$. Note that when we have perfect CSI, i.e. $\rho_{m,k} = 0$, (4.16) actually reduces to the multi-level waterfilling equation for perfect CSI in (4.30). Unlike (3.5), there is no closed form solution to (4.16), but it can be solved using numerical integration of the expectation, and a zero-finding procedure like bisection method [66] to find the power allocation.

Plugging (4.16) into (4.15b) and then in (4.14), we have

$$g^* = \min_{\lambda \geq 0} \lambda \bar{P} + \sum_{k \in \mathcal{K}} \max_{m \in \mathcal{M}} \left(w_m \mathbb{E}_{\gamma_{m,k}} \left\{ R_{m,k} \left(\tilde{p}_{m,k}(\lambda) \gamma_{m,k} \right) | \hat{\gamma}_{m,k} \right\} - \lambda \tilde{p}_{m,k}(\lambda) \right)$$
$$\tag{4.17}$$

Using standard duality arguments (see e.g. [57, Prop. 5.1.2]), the objective in (4.17) can be shown to be convex in the single variable λ, but is actually not continuously differentiable due to the presence of the max function. Hence, powerful derivative-based minimization methods such as Newton's method cannot be used. Fortunately, similar to the perfect CSI case of Chapter 3, we can use derivative-free single-dimensional line search methods that only need function evaluations, e.g. Golden-section or Fibonacci search [66], to find the optimum multiplier λ^*.

4.3.3 Power Allocation Function Approximation

Although tractable, solving (4.17) is still highly computationally intensive, since for each candidate λ in the line search iterations, we need to compute MK power allocation values (4.16), each of which requires a zero-finding routine where a function value evaluation involves numerical integration to compute the expectation. Although both the line search and the zero-finding routines typically converge within very few iterations (< 10 in our experiments), the numerical integration procedure requires a lot more iterations (> 50), and hence is the primary computational bottleneck. We shall overcome this problem using a closed-form approximation to the expectation in the power allocation function (4.16).

The Gamma distribution is known to approximate the body of the Chi-squared pdf quite well [11, p. 55]. Although the Gamma distribution approximation is poor for the tail of the Chi-squared pdf, it does not affect us very much, since we use the pdf to take the expectation, and thus the small values of the tails do not affect the approximation too adversely (the approximation is shown in Fig. 4.1). Thus, it is possible to use a Gamma distribution to approximate the NCχ^2 distribution of $\gamma_{m,k}|\hat{\gamma}_{m,k}$ (4.12)

$$f_{\gamma_{m,k}}(\gamma_{m,k}|\hat{\gamma}_{m,k}) \approx \frac{\beta^\alpha}{\Gamma(\alpha)}\gamma_{m,k}^{\alpha-1}e^{-\beta\gamma_{m,k}} \qquad (4.18)$$

where $\alpha = (\kappa_{m,k}+1)^2/(2\kappa_{m,k}+1)$ is the Gamma pdf shape parameter with $\kappa_{m,k} = \frac{\hat{\gamma}_{m,k}}{\rho_{m,k}}$ as the specular to diffuse power ratio, equivalent to the $K-$factor in a Ricean pdf; and $\beta = \alpha/(\hat{\gamma}_{m,k}+\rho_{m,k})$ is the Gamma pdf rate parameter.

Using this pdf, we can use [89, Section 3.383.10] to arrive at the following closed form approximation to the integral

$$\mathbb{E}_{\gamma_{m,k}}\left\{\frac{\gamma_{m,k}}{1+p_{m,k}\gamma_{m,k}}\bigg|\hat{\gamma}_{m,k}\right\} \approx \frac{\beta^\alpha}{\Gamma(\alpha)}\int_0^\infty \frac{\gamma_{m,k}^\alpha}{1+p_{m,k}\gamma_{m,k}}e^{-\beta\gamma_{m,k}}d\gamma_{m,k}$$
$$= \frac{\alpha}{p_{m,k}}\left(\frac{\beta}{p_{m,k}}\right)^\alpha e^{\frac{\beta}{p_{m,k}}}\Gamma\left(-\alpha,\frac{\beta}{p_{m,k}}\right) \qquad (4.19)$$

where $\Gamma(a,x)$ is the incomplete Gamma function [89, Section 8.350]. Using (4.19) in (4.16) to solve for $p_{m,k}$, we are able to closely approximate the power allocation function $\tilde{p}_{m,k}$. We plot the power allocation function using the Gamma pdf approximation and the actual Chi-squared pdf in Fig. 4.1 with $\gamma_0 = 1$ for various $\rho_{m,k} = \hat{\sigma}_{m,k}^2/\sigma_\nu^2$. Note that the approximation error is negligible, with a normalized mean-squared error of 5×10^{-5} and maximum

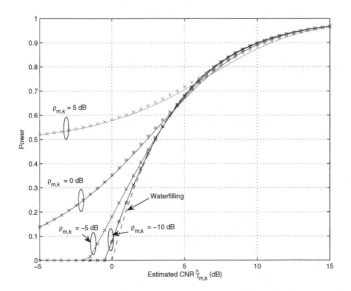

Fig. 4.1. Power allocation as a function of estimated CNR $(\hat{\gamma})$ with $\gamma_0 = 1$ for various $\rho_{m,k} = \hat{\sigma}_{m,k}^2/\sigma_{\nu}^2$, where '$-$' is the optimal power allocation, '$\times$' the approximation, and '$--$' the waterfilling solution for perfect CSI ($\rho_{m,k} = -\infty$ dB).

error of 2.7×10^{-4}, while the computation of the approximation is almost $300\times$ faster than direct numerical integration using very crude computational time measurements in Matlab 7.2 (`tic-toc`).

4.3.4 Optimal Subcarrier and Power Allocation

The optimal multiplier λ^* determines the optimal cutoff SNR $\gamma_{0,m}^* = \frac{\lambda^* \ln 2}{w_m}$, which in turn determines (4.16) to arrive at the optimal user selection and power allocation per subcarrier k:

$$m_k^* = \arg \max_{m \in \mathcal{M}} \mathbb{E}_{\gamma_{m,k}} \left\{ w_m R_{m,k} \left(\tilde{p}_{m,k}(\lambda^*) \gamma_{m,k} \right) | \hat{\gamma}_{m,k} \right\} - \lambda^* \tilde{p}_{m,k}(\lambda^*) \quad (4.20)$$

$$p_{m,k}^* = \tilde{p}_{m,k}(\lambda^*)\mathbf{1}(m = m_k^*) \quad (4.21)$$

Fig. 4.2 presents a flow chart of the OFDMA weighted-sum continuous rate maximization algorithm with partial CSI.

Note that, similar to the perfect CSI case, it is possible that the candidate power allocation values do not satisfy the total power constraint, since this constraint is not enforced explicitly. We use a similar heuristic of scaling the final power allocation values by $\bar{P}/\hat{P}_{tot}(\lambda^*)$ where

$$\hat{P}_{tot}(\lambda^*) = \sum_{k \in \mathcal{K}} \tilde{p}_{m_k^*,k}(\lambda^*) \quad (4.22)$$

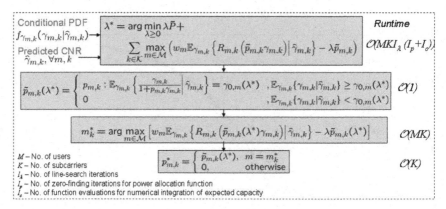

Fig. 4.2. OFDMA resource allocation algorithm for weighted-sum continuous rate maximization with partial CSI.

is the total power. Using (4.22) in (4.13), we arrive at our computed feasible primal optimal value

$$\hat{f}^* = \sum_{k\in\mathcal{K}} w_{m_k^*}\mathbb{E}_{\gamma_{m_k^*,k}}\left\{\log_2\left(1 + \frac{\bar{P}}{\hat{P}_{tot}(\lambda^*)}\tilde{p}_{m_k^*,k}(\lambda^*)\gamma_{m_k^*,k}\right)\bigg|\hat{\gamma}_{m_k^*,k}\right\} \quad (4.23)$$

4.3.5 Complexity Analysis

In analyzing the complexity, note that in each search iteration for λ in (4.17), we need to compute MK candidate power allocation functions given by (4.16) and (4.19). Each power allocation value calculation requires a zero-finding routine, e.g. bisection or Newton search [66], which we assume requires I_p function evaluations to converge. After determining the power allocation value, we then use it in the ergodic capacity integral in (4.17), which we assume requires I_c function evaluations to compute. Finally, assuming that we require I_λ line search iterations to solve for the optimum λ, the overall complexity is $\mathcal{O}(I_\lambda MK(I_p + I_c))$. Ignoring the constants I_λ, I_p, and I_c, the complexity is just $\mathcal{O}(MK)$.

4.4 Discrete Rate Maximization with Partial CSI and CDI

Consider the discrete rate function given in (2.8) using a slightly different convention in terms of the cut-off SNR indices for notational convenience in the subsequent development

$$R_{m,k}^d(p_{m,k}\gamma_{m,k}) = \begin{cases} 0, & p_{m,k}\gamma_{m,k} < \eta_0 \\ r_1, & \eta_0 \le p_{m,k}\gamma_{m,k} < \eta_1 \\ \vdots & \vdots \\ r_L, & \eta_{L-1} \le p_{m,k}\gamma_{m,k} < \eta_L \equiv \infty \end{cases} \qquad (4.24)$$

where $\{r_l\}_{l \in \mathcal{L}}$, $\mathcal{L} = \{1, \ldots, L\}$ are the L available discrete information rates in increasing order, and $\{\eta_l\}_{l=0}^{L-1}$ are the SNR boundaries chosen in such a way that the information rate r_l is supportable subject to an instantaneous BER constraint.

In the perfect CSI case of Sec. 3.3, the candidate power allocation function that satisfies the BER constraint for each possible rate r_l is simply multi-level fading inversion (MFI), i.e. $p_{m,k}^{(l)} = \eta_l/\gamma_{m,k}$. This allows us to do away with explicitly imposing the BER constraint, since all that we require are the SNR rate region boundaries η_l which can be computed offline. However, with imperfect CSI, the average rate is given as

$$\bar{R}_{m,k} = \sum_{l \in \mathcal{L}} r_l P\left(\eta_{l-1} \le p_{m,k}\gamma_{m,k} < \eta_l | \hat{\gamma}_{m,k}\right) \qquad (4.25)$$

Since we do not have the perfect CSI information $\gamma_{m,k}$, simply performing MFI on the imperfect CSI $\hat{\gamma}_{m,k}$ does not guarantee satisfaction of the BER constraint, and is illustrated in the results section (Section 4.5). This necessitates a different approach for fulfilling the BER constraint.

4.4.1 Closed-form average BER function

With the imperfect CSI assumption, we require a BER function that can be expressed in terms of the SNR $p_{m,k}\gamma_{m,k}$ for a given r_l in order to enforce the *average* BER constraint. Suppose that we have this BER function for a given rate r_l denoted as $\text{BER}_l(p_{m,k}\gamma_{m,k})$, which could be derived using theoretical analysis or curve fitting from empirical data, the average BER constraint can be written as

$$\mathbb{E}_{\gamma_{m,k}}\left\{\text{BER}_l(p_{m,k}\gamma_{m,k})|\hat{\gamma}_{m,k}\right\} = \overline{\text{BER}} \qquad (4.26)$$

Solving for $p_{m,k}$ in (4.26) for each $l \in \mathcal{L}$, we have L power allocation functions to choose from.

In order to simplify our development, we derive a closed-form expression for (4.26) assuming the fading distributions derived in Section 4.2, and a representative BER prototype function that has been empirically shown to fit a lot of practical scenarios (see e.g. [24]). This prototype BER function is given by

$$\text{BER}_l(p_{m,k}\gamma_{m,k}) = a_l e^{-b_l p_{m,k}\gamma_{m,k}} \qquad (4.27)$$

where a_l and b_l are constants that are searched to fit the actual BER function for each r_l. For example, if we assume a Grey-coded square 2^{r_l}-QAM modulation scheme in AWGN, the BER function can be approximated to within

1-dB for $r_l \geq 2$ and BER $\leq 10^{-3}$ with $a_l = 0.2$ and $b_l = 1.6/(2^{r_l} - 1)$ [24]. Using (4.27) in (4.26) with the pdf in (4.12), we have

$$\mathbb{E}_{\gamma_{m,k}}\{\mathrm{BER}_l(p_{m,k}\gamma_{m,k})|\hat{\gamma}_{m,k}\} = \tilde{a}_{m,k}^{(l)} \int_0^\infty e^{-x\left(\tilde{b}_{m,k}^{(l)}p_{m,k}+1\right)} I_0\left(2\sqrt{\kappa_{m,k}x}\right) dx \tag{4.28}$$

where $x = \frac{\gamma_{m,k}}{\rho_{m,k}}$, $\tilde{a}_{m,k}^{(l)} = a_l e^{(-\kappa_{m,k})}$, $\tilde{b}_{m,k}^{(l)} = b_l \rho_{m,k}$, and $\kappa_{m,k} = \frac{\hat{\gamma}_{m,k}}{\rho_{m,k}}$.

Note that (4.28) can be interpreted as the Laplace transform of $I_0(2\sqrt{\kappa_{m,k}x})$ with parameter $s = \tilde{b}_{m,k}^{(l)}p_{m,k} + 1$, which is given in [90, Eq. 29.3.81]. Hence, (4.28) can be written as

$$\mathbb{E}_{\gamma_{m,k}}\{\mathrm{BER}_l(p_{m,k}\gamma_{m,k})|\hat{\gamma}_{m,k}\} = \frac{\tilde{a}_{m,k}^{(l)}}{\tilde{b}_{m,k}^{(l)}p_{m,k}+1} e^{\frac{\kappa_{m,k}}{\tilde{b}_{m,k}^{(l)}p_{m,k}+1}} \tag{4.29}$$

4.4.2 Closed-form power allocation function

Equating (4.29) with the target $\overline{\mathrm{BER}}$, we arrive at the closed form expression for the candidate power allocation function given the estimated CNR $\hat{\gamma}_{m,k}$ and data rate r_l (see Appendix E for derivation)

$$\tilde{p}_{m,k}^{(l)} = \frac{1}{\tilde{b}_{m,k}^{(l)}} \left(\frac{\kappa_{m,k}}{W\left(\overline{\mathrm{BER}}\,\kappa_{m,k}/\tilde{a}_{m,k}^{(l)}\right)} - 1 \right) \tag{4.30}$$

where $W(x)$ is the *Lambert-W* function (see (3.10) for the Lambert-W function used in a different context.). It is important to emphasize that (4.30) gives us the power allocation value that fulfills the average BER constraint when r_l is chosen as the rate given imperfect CSI $\hat{\gamma}_{m,k}$. Fig. 4.3 shows the power allocation as a function of the estimated CNR $\hat{\gamma}_{m,k}$ for uncoded 4−QAM and 64−QAM for various $\rho_{m,k}$. We also plot the power allocation function when treating the $\hat{\gamma}_{m,k}$ as perfect, i.e. $p_{m,k}^{(l)} = \eta_l/\hat{\gamma}_{m,k}$ called multi-level fading inversion on imperfect CSI (Imperfect CSI-MFI). We can see that as $\rho_{m,k}$ decreases (prediction accuracy increases), the power allocation function approaches that of Imperfect CSI-MFI. On the other hand, a higher $\rho_{m,k}$ requires higher power in order to ensure the average BER requirement is met, esp. for low $\hat{\gamma}_{m,k}$. Note also that the power allocation functions approach the Imperfect CSI-MFI value as $\hat{\gamma}_{m,k}$ becomes large, despite the value of $\rho_{m,k}$.

4.4.3 Closed-form average rate function

Using (4.30) in (4.25), the average rate given that r_l is chosen as the transmission rate can be written as

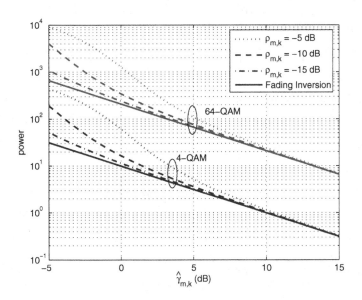

Fig. 4.3. Discrete rate power allocation as a function of estimated CNR ($\hat{\gamma}$) with $\gamma_0 = 1$ for various $\rho_{m,k}$.

$$
\begin{aligned}
\bar{R}_{m,k}(r_l) &= \sum_{i \in \mathcal{L}} r_i P\left(\eta_{i-1} \leq \tilde{p}_{m,k}^{(l)} \gamma_{m,k} < \eta_i | \hat{\gamma}_{m,k}\right) \\
&= \sum_{i \in \mathcal{L}} r_i P\left(\frac{\eta_{i-1}}{\tilde{p}_{m,k}^{(l)}} \leq \gamma_{m,k} < \frac{\eta_i}{\tilde{p}_{m,k}^{(l)}} \middle| \hat{\gamma}_{m,k}\right) \\
&= \sum_{i \in \mathcal{L}} r_i \left(F_{\gamma_{m,k}}\left(\frac{\eta_i}{\tilde{p}_{m,k}^{(l)}} \middle| \hat{\gamma}_{m,k}\right) - F_{\gamma_{m,k}}\left(\frac{\eta_{i-1}}{\tilde{p}_{m,k}^{(l)}} \middle| \hat{\gamma}_{m,k}\right) \right)
\end{aligned}
\tag{4.31}
$$

From [88, Eq. 2.1-124], we have the following closed-form expression for the cumulative distribution function (cdf) of a non-central Chi-squared random variable

$$
F_{\gamma_{m,k}}\left(x | \hat{\gamma}_{m,k}\right) = 1 - Q\left(\sqrt{\frac{2\hat{\gamma}_{m,k}}{\rho_{m,k}}}, \sqrt{\frac{2x}{\rho_{m,k}}}\right)
\tag{4.32}
$$

where $Q(a,b)$ is the Marcum-Q function [88, Eq. 2.1-122]. Using (4.32) in (4.31), we have a closed-form expression for the average rate for user m and subcarrier k given a choice of transmission rate r_l.

4.4.4 Problem Formulation

Considering the above development, we can think of our decision variables in this case as a vector of rate allocation indices $\boldsymbol{l} = [\boldsymbol{l}_1^T, \ldots, \boldsymbol{l}_K^T]^T$ where $\boldsymbol{l}_k^T =$

$[l_{1,k}, \ldots, l_{M,k}]^T$ and $l_{m,k} \in \{0, 1, \ldots, L\}$. The exclusive subcarrier assignment restriction can be written as $l_k \in \mathcal{L}_k$, where

$$\mathcal{L}_k = \{l_{m,k} \in \{0, 1, \ldots, L\} | l_{m,k} l_{m',k} = 0; \forall m \neq m'; m, m' \in \mathcal{M}\} \quad (4.33)$$

For notational convenience, we let $l \in \mathcal{L} = \mathcal{L}_1 \times \cdots \times \mathcal{L}_K$ denote the space of allowable rate allocation indices for all subcarriers. Note that a decision of $l_{m,k} = 0$ means that neither rate nor power is transmitted on subcarrier k by user m. Thus, we can define $\bar{R}_{m,k}(r_0) \equiv 0$ and $\tilde{p}_{m,k}^{(0)} \equiv 0$. The discrete weighted sum rate maximization problem with partial CSI is then formulated as

$$
\begin{aligned}
f_d^* = \max_{l \in \mathcal{L}} \quad & \sum_{m \in \mathcal{M}} w_m \sum_{k \in \mathcal{K}} \bar{R}_{m,k}(r_{l_{m,k}}) \\
\text{s.t.} \quad & \sum_{m \in \mathcal{M}} \sum_{k \in \mathcal{K}} \tilde{p}_{m,k}^{(l_{m,k})} \leq \bar{P}
\end{aligned}
\quad (4.34)
$$

where the power allocation function is given by (4.30) and the average rate by (4.31).

4.4.5 Dual Optimization Framework

Following a similar development as in Section 4.3, the dual problem can be written as

$$g_d^* = \min_{\lambda \geq 0} \lambda \bar{P} + \sum_{k \in \mathcal{K}} \max_{m \in \mathcal{M}} \max_{l \in \mathcal{L} \cup \{0\}} \left(\bar{R}_{m,k}(r_l) - \lambda \tilde{p}_{m,k}^{(l)} \right) \quad (4.35)$$

where we can once again use a univariate line-search method such as Golden-section search to compute for the optimum multiplier λ_d^*. Note that neither (4.30) nor (4.31) depend on λ. Hence, we can pre-compute these quantities for all $l \in \mathcal{L}$, $m \in \mathcal{M}$, and $k \in \mathcal{K}$ before running the line search iterations. Using λ_d^*, we arrive at the optimal rate allocation indices

$$l_{m,k}^* = \arg \max_{l \in \mathcal{L}} w_m \bar{R}_{m,k}(r_l) - \lambda_d^* \tilde{p}_{m,k}^{(l)} \quad (4.36)$$

which in turn give us the optimal subcarrier, rate, and power allocation:

$$m_k^* = \arg \max_{m \in \mathcal{M}} w_m \bar{R}_{m,k}(r_{l_{m,k}^*}) - \lambda_d^* \tilde{p}_{m,k}^{(l_{m,k}^*)} \quad (4.37)$$

$$p_{m,k}^* = \tilde{p}_{m,k}^{(l_{m,k}^*)} \mathbf{1}(m = m_k^*) \quad (4.38)$$

$$r_{m,k}^* = r_{l_{m,k}^*} \mathbf{1}(m = m_k^*) \quad (4.39)$$

Finally, similar to the perfect CSI continuous rate case in Section 3.2.8, the duality gap can be computed as in (3.22) to characterize how far away the solution is from the optimal. Fig. 4.4 presents a flow chart of the OFDMA weighted-sum discrete rate maximization algorithm with partial CSI.

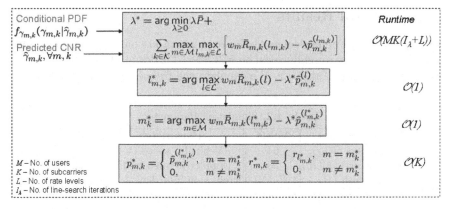

Fig. 4.4. OFDMA resource allocation algorithm for weighted-sum discrete rate maximization with partial CSI.

4.4.6 Complexity Analysis

Before running the line search iterations to compute for λ^* in (4.35), we need to compute MKL power allocation values (4.30) and average rate values (4.31) and store it in memory. This is followed by the search iterations which we assume to require I_λ, wherein each iteration requires $\mathcal{O}(MK)$ operations (4.35). The overall complexity order for the discrete rate resource allocation algorithm is thus $\mathcal{O}(MK(L + I_\lambda))$. Since L and I_λ are just constants independent of M and K, the complexity is $\mathcal{O}(MK)$.

Table 4.1. Comparison of the proposed continuous and discrete rate resource allocation algorithms assuming partial CSI, with the suboptimal method of using the perfect CSI algorithms on the imperfect CSI.

Algorithm	Per-symbol Complexity	Data Rate	Ave. BER	Rel. Gap Order
Cont. Rate, Proposed	$\mathcal{O}(MKI_\lambda(I_p + I_c))$	High	−	10^{-6}
Cont. Rate, MWF	$\mathcal{O}(I_\lambda MK)$	High	−	−
Disc. Rate, Proposed	$\mathcal{O}(MK(I_\lambda + L))$	Med.	1.0×10^{-3}	10^{-4}
Disc. Rate, MFI	$\mathcal{O}(I_\lambda MK \log L)$	Med.	1.8×10^{-3}	−

$\overline{\text{BER}} = 10^{-3}$, M-no. of users, K-no. of subcarriers, I_λ-no. of line search iterations for dual problem, I_p-no. of zero-finding iterations for the power allocation function (4.16), I_c-no. of function evaluations for numerical integration of the expected capacity (4.17),L-no. of discrete rate levels (2.8)

4.5 Numerical Results

We present several numerical examples to substantiate our theoretical claims. Our simulations are roughly based on a 3GPP-LTE downlink [19] system with parameters given in Table 4.2. We simulate the frequency-selective Rayleigh

Table 4.2. Simulation Parameters

Parameter	Value	Parameter	Value
Subcarriers (K_{fft})	64	Vehicular speed (V)	120 kph
Used Subcarriers (K)	33	Doppler frequency (F_d)	289 Hz
Bandwidth (B)	1.25 MHz	Prediction filter length (P)	4
Sampling Freq. (F_s)	1.92 MHz	Pilot spacing (D_t)	7
Carrier Freq. (F_c)	2.6 GHz	CP Length (L_{cp})	6 samples

fading channel using the ITU-Vehicular A channel model [71]. We assume Clarke's U-shaped power spectrum [11] for each multipath tap, resulting in the temporal autocorrelation function $r_m[\Delta] = J_0(2\pi\Delta F_d D_t (K_{\text{fft}} + L_{cp})/F_s)$ where $J_0(x)$ is the zeroth-order Bessel function of the first kind [90, Ch. 9]. To simulate imperfect CSI, we generate IID realizations of $\hat{\boldsymbol{h}}_m$ and its prediction error vector $\hat{\boldsymbol{e}}_m$ given in (4.11). This allows us to also generate the "actual" channel $\boldsymbol{h}_m$ for the perfect CSI cases using (4.11).

4.5.1 Continuous Rate Case

In Fig. 4.5, we show the two-user capacity region for continuous rate allocation with imperfect CSI (Imperfect CSI-Optimal) with 5000 channel realizations per data point. We also show the capacity region using optimal instantaneous rate resource allocation assuming perfect CSI (Perfect CSI-Optimal), which is essentially multi-level waterfilling (MWF) (see Sec. 3.2.6); and the capacity region when we simply use MWF on the imperfect CSI (Imperfect CSI-MWF). Note that in all cases, rate maximization with imperfect CSI through channel prediction performs quite close to the case with perfect CSI. More important, Imperfect CSI-MWF performs similar to Imperfect CSI-Optimal. This can be explained by noticing that the optimal power allocation assuming imperfect CSI is almost equal to the waterfilling curve (see Fig. 4.1) except for very low estimated CNR. However, due to the effect of frequency and multiuser diversity, the subcarrier is typically assigned to the user with the highest CNR; thus, the power allocation is quite often almost identical to performing waterfilling on the imperfect CSI. A similar observation was also made in [84], albeit for the single user scenario.

4.5.2 Discrete Rate Case

Fig. 4.6 shows the discrete rate region for the optimal resource allocation algorithm assuming imperfect CSI (Imperfect CSI-Optimal). We also show the

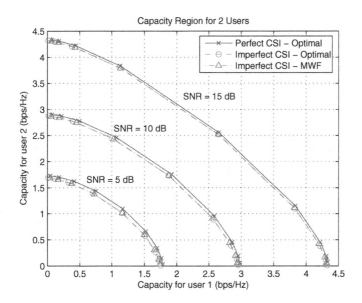

Fig. 4.5. Two-user capacity region for continuous rate optimal resource allocation with imperfect CSI. We also show the capacity region for optimal allocation with perfect CSI, and using multi-level waterfilling (MWF) on the imperfect CSI.

rate region achieved by using optimal resource allocation for discrete rates with perfect CSI (Perfect CSI-Optimal), which is essentially MFI (see Sec. 4.4.5), and by using MFI on the imperfect CSI (Imperfect CSI-MFI). Observe that due to the imperfect CSI assumption, Imperfect CSI-Optimal loses approximately 8% of the sum capacity when compared to Perfect CSI-Optimal. Observe also that Imperfect CSI-MFI results in a rate region that is quite close to the Perfect CSI-Optimal, and actually results in higher raw rates than the Imperfect CSI-Optimal. However, if we consider the average BER for each subcarrier shown in Fig. 4.7, Imperfect CSI-Optimal actually meets the average BER constraint of 10^{-3} (within $\pm 2\%$), but Imperfect CSI-MFI results in average BER violations of between $30 - 180\%$. Interestingly, the shape of the BER for Imperfect-CSI-Suboptimal closely resembles the shape of the prediction error variance $\hat{\sigma}_{m,k}^2$, shown in Fig. 4.8. This is intuitively satisfying, since a larger prediction error results in a larger mismatch between perfect and imperfect CSI, which is not taken into account by the Imperfect CSI-MFI algorithm. Thus, Imperfect CSI-MFI is equally aggressive in rate and power allocation even when the CSI prediction error is quite large. Our proposed Imperfect CSI-Optimal algorithm, on the other hand, is actually more conservative in rate and power allocation when the prediction MSE is large, thus allowing the average BER to be met. In a practical communications system, this would mean the difference of whether a packet is decoded successfully or not. Thus, using Imperfect CSI-MFI would result in unnecessary

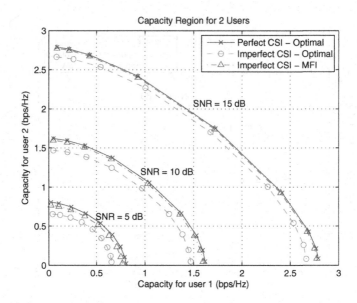

Fig. 4.6. Two-user capacity region for discrete rate optimal resource allocation with imperfect CSI. We also show the capacity region for optimal allocation with perfect CSI, and using multi-level fading inversion (MFI) on the imperfect CSI.

packet retransmissions and delays, and consequently decrease the throughput significantly. An explicit characterization in terms of throughput, however, is beyond the scope of this book.

Table 4.3 shows the other relevant metrics of the optimal resource algorithms. The first column shows the average number of line-search iterations it took to converge to a tolerance of 10^{-4}. The second column shows the resulting relative duality gaps. We can see that the duality gaps are virtually zero, and thus both algorithms can be considered practically optimal for this set of simulation parameters.

Table 4.3. Relevant Performance Measures for the Proposed Resource Allocation Algorithms

Metric	[a]No. of Iterations (I_λ)			[b]Relative Gap ($\times 10^{-4}$)		
SNR	5 dB	10 dB	15 dB	5 dB	10 dB	15 dB
Continuous Rates	8.599	8.501	8.686	.0840	.0568	.0412
Discrete Rates	21.33	21.15	21.12	71.48	7.707	5.662

[a] Average no. of line search iterations to solve the dual problem in (4.17) or (4.35)

[b] Average relative duality/optimality gap given in (3.22) but for the partial CSI case

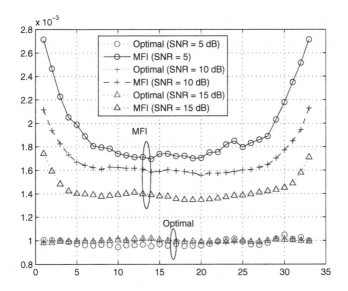

Fig. 4.7. Average BER for both users in each subcarrier for Imperfect CSI-Optimal and Imperfect CSI-MFI.

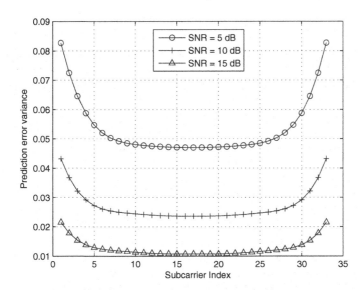

Fig. 4.8. Prediction error variance $(\hat{\sigma}_{m,k}^2)$ for each subcarrier for different SNRs.

4.5.3 Complexity Comparison

Table 4.1 summarizes the complexity analysis for both continuous and discrete rate algorithms. Note that the ability to pre-compute the power and rate allocations, and the existence of closed-form solutions for these in the discrete rate allocation case in contrast to the continuous rate case makes discrete rate allocation less complex than the continuous rate allocation. This is fortunate because the discrete rate case is more practically relevant. Note that this is reverse of what we observed for the perfect CSI case in Chapter 3.

4.6 Conclusion

We have derived optimal resource allocation algorithms for ergodic continuous and discrete rate maximization in OFDMA downlinks assuming the availability of partial CSI. Using a dual optimization approach, we derived algorithms with complexity $\mathcal{O}(MK)$ per iteration and achieve relative duality gaps that are less than 10^{-5} for continuous rates and 10^{-3} for discrete rates in typical scenarios. Although the solution framework of the imperfect CSI case is similar to the perfect CSI case in Ch. 3, the solution derivations, discussions, and complexity issues are quite different.

The important conclusions/contributions of this chapter are:

- *Partial CSI assumption disallows averaging across temporal dimension:* In the perfect CSI case, we can assume that the power allocation vector is a function of the perfectly known channel gains. Hence, assuming ergodicity of the channel gains, we are able to exploit the temporal dimension by imposing an average power constraint and allowing the *total power* in each time instance to vary, as long as the *average power* over time is met, giving us an additional degree of freedom to exploit. Unfortunately, this is not possible in the imperfect CSI case, because we only have information on the estimated channel gains, and it is difficult to assume ergodicity of the estimated channel information.

- *Closed-form approximation for the power allocation function in continuous rate case:* In the perfect CSI case, the per-user per-subcarrier power allocation is given neatly by the closed-form multi-level waterfilling solution (3.5). In the imperfect CSI case considered in this chapter, there is no closed-form solution for the power allocation function given in (4.16), thus, the approximate closed-form power allocation function with negligible approximation error is crucial to our development.

- *Derivation of closed-form average BER, power allocation, and average rate functions in the discrete rate case:* Assuming perfect CSI, the power allocation function that fulfills the BER constraint for each discrete rate level is a straightforward multilevel fading inversion, since the transitions in the staircase discrete rate function are where the BER is met with the least power for each rate level (3.33). Consequently, the rate allocation is likewise straightforward to compute. In the imperfect CSI case, it is not as simple, since we only have information on the estimated channel gain $\hat{\gamma}_{m,k}$ and the quality of the estimate $\rho_{m,k}$. Thus, it is crucial to the imperfect CSI case that the average BER function (4.28), the power allocation function that fulfills the average BER (4.30), and the average rate function resulting from an instantaneous rate allocation decision (4.31) are derived. Surprisingly, we were able to express all of these as functions of $\hat{\gamma}_{m,k}$ and $\rho_{m,k}$ in closed-form, allowing us to reformulate the problem into a deterministic integer program with a separable objective function (4.34). This reformulation is novel, and it made the dual optimization method feasible in our case.

5

Rate Maximization with Proportional Rate Constraints

5.1 Introduction

In the previous two chapters, we did not impose any constraints on the data rate of the users, and fairness is assumed to be imposed by the weights w_m. In some cases, imposing ratio constraints among the users' rates is more useful [45] [91] [46] [92], i.e.

$$R_1 : \ldots : R_M = \phi_1 : \ldots : \phi_M \tag{5.1}$$

where R_m is the rate of user m, and the ϕ_m terms are the given proportionality constants which we can assume to satisfy $\sum_{m \in \mathcal{M}} \phi_m = 1$. Sum rate maximization subject to this proportional rates constraint allows a more definitive prioritization among the users, which is quite useful for service class differentiation. Theoretically, this formulation also traces out the boundary of the capacity region similar to the weighted sum-rate maximization. The main difference is that it actually identifies the points on the capacity region boundary that satisfy the rate proportionality constraints. Furthermore, the max-min rate formulation is a special case of this formulation, i.e. when $\phi_1 = \ldots = \phi_M$.

The instantaneous sum-rate maximization with proportional rate constraints have been studied previously in [45] [91] [46]. The main emphasis of these papers, in terms of formulation, was on an instantaneous rate maximization with instantaneous proportional rate constraints. Furthermore, the solution methods proposed were suboptimal heuristics with complexity of $\mathcal{O}(MK \log(K))$ or higher.

In this chapter, we use a similar dual optimization framework to solve the *ergodic sum-rate* maximization with *proportional ergodic rate* constraints. We show that the proportional rate constraints can actually be imposed by a similar weighted-sum rate dual, with the weights being the *dual optimal geometric multipliers* themselves that enforce the proportional rate constraints. Thus, we can use the techniques in Chapters 3-4 with the additional operation of determining the optimal weights. We emphasize the continuous rate, perfect CSI formulation in this chapter, but extensions to the discrete rate and/or partial CSI assumptions are shown to be straightforward extensions. We compared the performance of our algorithm with the previous algorithm that gives the best performance [46], and show that exploiting the temporal dimension using the ergodic formulation provides huge rate gains versus the previous state-of-the-art.

One main disadvantage of considering ergodic rates is the assumption that the *channel distribution information* (CDI) is perfectly known at the transmitter, and thus the expected values of the rates can be computed. Although methods to estimate the distribution function are available [93], they are typically more suitable for off-line processing rather than the online algorithms that are needed in practical wireless system implementations. Therefore, we also propose an adaptive algorithm based on *stochastic approximation* methods [94] [95] that do not require knowledge of the CDI, and is shown to converge to the optimal solution w.p.1, while requiring only $\mathcal{O}(MK)$ complexity per-symbol *without iterations*, since the iterations are actually done across time. Since the weighted-sum rate formulations in Chapters 3-4 are just special cases of the proportional rates case, these formulations can also be solved using this adaptive framework.

This chapter is organized as follows. Section 5.2 discusses the algorithm assuming perfect CSI, perfect CDI, and continuous rates. Section 5.3 relaxes the perfect CDI assumption and derives the adaptive algorithm that is shown to converge to the optimal rates. Section 5.4 presents numerical results and we conclude the chapter in Section 5.5.

5.2 Proportional Rate Maximization with Perfect CSI and CDI

5.2.1 Problem Formulation

The ergodic rate maximization problem with proportional ergodic rate constraints can be formulated as

$$
\max_{p(\gamma)\in\mathcal{P}} \mathbb{E}_\gamma \left\{ \sum_{m\in\mathcal{M}} \sum_{k\in\mathcal{K}} R_{m,k}(p_{m,k}\gamma_{m,k}) \right\}
$$

$$
\text{s.t. } \mathbb{E}_\gamma \left\{ \sum_{m\in\mathcal{M}} \sum_{k\in\mathcal{K}} p_{m,k} \right\} \leq \bar{P}
$$

$$
\mathbb{E}_\gamma \left\{ \sum_{k\in\mathcal{K}} R_{m,k}(p_{m,k}\gamma_{m,k}) \right\} \geq \phi_m \mathbb{E}_\gamma \left\{ \sum_{m\in\mathcal{M}} \sum_{k\in\mathcal{K}} R_{m,k}(p_{m,k}\gamma_{m,k}) \right\}, \forall m \in \mathcal{M}
$$

$$(5.2)$$

where the ϕ_m terms are the proportionality constants for each user m such that $\sum_m \phi_m = 1$. The constants ϕ_m can be interpreted as the portion of the total ergodic sum rate that should be allocated to each user m. We denote by $\boldsymbol{\phi} = [\phi_1, \ldots, \phi_M]^T$ the vector of proportionality constants. By introducing a dummy optimization variable $R \geq 0$ that represents the ergodic sum rate, we can rewrite the problem as

$$\max_{R \in \mathbb{R}^+, \boldsymbol{p}(\boldsymbol{\gamma}) \in \mathcal{P}} R$$

$$\text{s.t. } \mathbb{E}_{\boldsymbol{\gamma}} \left\{ \sum_{m \in \mathcal{M}} \sum_{k \in \mathcal{K}} p_{m,k} \right\} \leq \bar{P} \qquad (5.3)$$

$$\mathbb{E}_{\boldsymbol{\gamma}} \left\{ \sum_{k \in \mathcal{K}} R_{m,k}(p_{m,k}\gamma_{m,k}) \right\} \geq \phi_m R, \forall m \in \mathcal{M}$$

A similar reformulation as in (5.3) that uses a dummy variable is proposed in [41] to solve for the max-min rate, and in [92] for instantaneous proportional rates.

5.2.2 Dual Optimization Framework

The Lagrangian of (5.3) is given by

$$L(R, \boldsymbol{p}(\boldsymbol{\gamma}), \lambda, \boldsymbol{\mu}) = R + \lambda \left(\bar{P} - \mathbb{E}_{\boldsymbol{\gamma}} \left\{ \sum_{k \in \mathcal{K}} \sum_{m \in \mathcal{M}} p_{m,k} \right\} \right)$$

$$+ \sum_{m \in \mathcal{M}} \mu_m \left(\mathbb{E}_{\boldsymbol{\gamma}} \left\{ \sum_{k \in \mathcal{K}} R_{m,k}(p_{m,k}\gamma_{m,k}) \right\} - \phi_m R \right)$$

$$(5.4)$$

where $\boldsymbol{\mu} = [\mu_1, \ldots, \mu_M]^T$ is the vector of geometric multipliers that are used to enforce the proportionality constraints. The dual objective can then be written as

$$\Theta(\lambda, \boldsymbol{\mu}) = \max_{R \in \mathbb{R}^+, \boldsymbol{p}(\boldsymbol{\gamma}) \in \mathcal{P}} L(R, \boldsymbol{p}(\boldsymbol{\gamma}), \lambda, \boldsymbol{\mu})$$

$$= \lambda \bar{P} + \max_{R \in \mathbb{R}^+, \boldsymbol{p}(\boldsymbol{\gamma}) \in \mathcal{P}} R \left(1 - \boldsymbol{\mu}^T \boldsymbol{\phi} \right) - \lambda \mathbb{E}_{\boldsymbol{\gamma}} \left\{ \sum_{k \in \mathcal{K}} \sum_{m \in \mathcal{M}} p_{m,k} \right\}$$

$$+ \sum_{m \in \mathcal{M}} \mu_m \mathbb{E}_{\boldsymbol{\gamma}} \left\{ \sum_{k \in \mathcal{K}} R_{m,k}(p_{m,k}\gamma_{m,k}) \right\}$$

$$(5.5)$$

Focusing on the first term in the maximization $R \left(1 - \boldsymbol{\mu}^T \boldsymbol{\phi} \right)$, we observe that if $1 - \boldsymbol{\mu}^T \boldsymbol{\phi} > 0$, then the optimal solution would be $R^* = \infty$, since R is a free variable. This is clearly an infeasible solution for the ergodic sum rate. Furthermore, if $1 - \boldsymbol{\mu}^T \boldsymbol{\phi} < 0$, then the optimal solution would be $R^* = 0$. In this case, the ergodic sum rate is zero and is uninteresting from an optimization point of view. Thus, we would like to constrain the multipliers to satisfy $\boldsymbol{\mu}^T \boldsymbol{\phi} = 1$, which allows us to remove the dummy variable from consideration since $R\left(1 - \boldsymbol{\mu}^T \boldsymbol{\phi}\right) = 0$. Thus, following the development in (3.4a)-(3.4e), (5.5) can be simplified to

$$\Theta(\lambda, \boldsymbol{\mu}) = \lambda \bar{P} + K \mathbb{E}_{\gamma_k} \left\{ \max_{m \in \mathcal{M}} \left(\mu_m R_{m,k}(\tilde{p}_{m,k} \gamma_{m,k}) - \lambda \tilde{p}_{m,k} \right) \right\} \tag{5.6}$$

where $\tilde{p}_{m,k} = \left[\mu_m/(\lambda \ln 2) - 1/\gamma_{m,k} \right]^+$. The main difference is that the "weights" in this case are no longer pre-determined constants, but are effectively the multipliers μ_m that enforce the proportional rate constraints. The dual problem can then be written as

$$g^* = \min_{\lambda \geq 0, \boldsymbol{\mu} \in \mathcal{U}} \Theta(\lambda, \boldsymbol{\mu}) \tag{5.7}$$

where $\mathcal{U} = \left\{ \boldsymbol{\mu} \geq 0 \,\middle|\, \boldsymbol{\mu}^T \boldsymbol{\phi} = 1 \right\}$. Given a candidate feasible $\boldsymbol{\mu}$, we can proceed using line search methods similar to Sec. 3.2.2 to solve for a candidate $\lambda^*(\boldsymbol{\mu})$ that enforces the power constraint. Another possible method is to use subgradient search [57, Ch. 6.3.1], which is a generalization of gradient-based search methods to possibly non-differentiable functions. From an initial guess λ^0, the subgradient method generates a sequence of dual feasible points according to the iteration

$$\lambda^{i+1} = \left[\lambda^i - s^i g_\lambda^i \right]^+ \tag{5.8}$$

where g_λ^i denotes the subgradient of $\Theta(\lambda^*(\boldsymbol{\mu}^i), \boldsymbol{\mu}^i)$ with respect to λ, and s^i is a positive scalar step-size. A similar subgradient method can be used to search for the optimal $\boldsymbol{\mu}^*$ that enforces the proportional rate constraints, given by the iterations

$$\boldsymbol{\mu}^{i+1} = \Pi_{\mathcal{U}} \left[\boldsymbol{\mu}^i - s^i g_\mu^i \right] \tag{5.9}$$

where g_μ^i denotes the subgradient of $\Theta(\lambda^*(\boldsymbol{\mu}^i), \boldsymbol{\mu}^i)$ with respect to $\boldsymbol{\mu}$, and $\Pi_{\mathcal{U}}[\cdot]$ denotes projection onto the set $\mathcal{U}$.

The subgradient method is particularly attractive for solving the dual problem, since the inequality constraint evaluated at the optimal power vector for a given λ and $\boldsymbol{\mu}$ is itself the subgradient [57], i.e.

$$g_\lambda^i = \bar{P} - \hat{P}_{tot}^i \tag{5.10}$$

where

$$\hat{P}_{tot}^i = \mathbb{E}_\gamma \left\{ \sum_{m \in \mathcal{M}} \sum_{k \in \mathcal{K}} p_{m,k}^*(\lambda^i, \mu_m^i) \right\} \tag{5.11}$$

is the average power given $\lambda^i, \boldsymbol{\mu}^i$; and

$$g_\mu^i = \bar{R}^i - \phi \bar{R}^i \tag{5.12}$$

where $\bar{R}^i = [\bar{R}_1^i, \dots, \bar{R}_M^i]$ with

$$\bar{R}_m^i = \mathbb{E}_\gamma \left\{ \sum_{k \in \mathcal{K}} R_{m,k}(p_{m,k}^*(\lambda^i, \mu_m^i)\gamma_{m,k}) \right\} \tag{5.13}$$

as the vector of ergodic rates per user for a given μ^i, and

$$\bar{R}^i = \sum_{m \in \mathcal{M}} \bar{R}_m^i \tag{5.14}$$

is the ergodic sum rate. The optimal power given the current λ^i and μ^i, $p_{m,k}^*(\lambda^i, \mu_m^i)$, is similarly derived as in (3.14)-(3.15) with w_m replaced by μ_m^i. The projection operation can be simply performed by clipping and rescaling the new iterate such that it is non-negative and that it satisfies $\mu_i^T \phi = 1$. Hence, we can write (5.9) as

$$\mu^{i+1} = \frac{[\mu^i - s^i g^i]^+}{\phi^T [\mu^i - s^i g^i]^+} \tag{5.15}$$

where $[x]^+$ implements $\max(x_i, 0)$ for each element in the vector argument x. The convergence properties of (5.15) for different step-size selection rules have already been studied previously (see e.g. [57, Ch. 6.3.1]). In our numerical experiments, we use the simple diminishing step-size rule

$$s^i = \frac{\beta}{i + \alpha} \tag{5.16}$$

where α and β are suitably chosen positive constants, which satisfies $s^i \to 0$ (for convergence) and $\sum_{i=0}^{\infty} s^i = \infty$ (for allowing us to go "anywhere").

We can interpret the multiplier vector μ^i as a vector of priorities for the users, wherein we try to increase the priority of a user while it is still unable to get its allocated "portion of the pie" $\phi_m \bar{R}^i$. Upon convergence, we arrive at the optimal μ^* which is the vector of appropriate weights for each user such that the proportionality constraints are met, and its corresponding λ^* that enforces the average power constraint.

5.2.3 Computation of the Per-user Ergodic Rate

Computing the optimal λ^* for a given μ has already been discussed thoroughly in Sec. 3.2.3, where all the development follows in a straightforward manner by simply replacing the weights w_m by μ_m. However, the computation of the subgradient requires knowing the individual ergodic sum rates per user, which was not required in the weighted sum rate case in Chapter 3. In order to derive this, we revisit the expectation integral used in computing the dual objective in (3.13), which we rewrite here in its expanded form by using (3.9)

$$\bar{g}_k = \int_0^\infty g_k \left(\sum_{m \in \mathcal{M}} f_{g_{m,k}}(g_k) \prod_{m' \neq m} F_{g_{m',k}}(g_k) \right) dg_k$$

$$= \sum_{m \in \mathcal{M}} \int_0^\infty g_k f_{g_{m,k}}(g_k) \prod_{m' \neq m} F_{g_{m',k}}(g_k) dg_k \qquad (5.17)$$

where the second equality can be interpreted as the sum of the per-user expected dual functions. By performing the following change of variables to revert back to the CNR variables (see (3.8) and (3.10))

$$g_k = g_{m,k}(\gamma_{m,k}, \lambda)$$
$$\gamma_{m,k} = \check{\gamma}_{m,k}(g_k) \qquad (5.18)$$

and using (3.11)-(3.12), we can rewrite (5.17) as

$$\bar{g}_k = \sum_{m \in \mathcal{M}} \int_{\gamma_{0,m}}^\infty g_{m,k}(\gamma_{m,k}, \lambda) f_{\gamma_{m,k}}(\gamma_{m,k}) \prod_{m' \neq m} F_{\gamma_{m',k}}(\check{\gamma}_{m',k}(g_{m,k}(\gamma_{m,k}, \lambda))) d\gamma_{m,k}$$

$$(5.19)$$

Focusing on the pdf term $f_{\gamma_{m,k}}(\gamma_{m,k}) \prod_{m' \neq m} F_{\gamma_{m',k}}(\check{\gamma}_{m',k}(g_{m,k}(\gamma_{m,k}, \lambda)))$, we can see an intuitive interpretation. This term can be interpreted as the joint pdf of $\gamma_{m,k}$ and the event that "user m wins subcarrier k." This is because the first term is the pdf of CNR for user m at subcarrier k, and the product terms can be seen as the probability that all other users m' for subcarrier k have CNRs less than $\check{\gamma}_{m',k}(g_{m,k}(\gamma_{m,k}, \lambda))$, which is equal to the probability that the marginal dual functions of all other users are less than that of user m, i.e. $\Pr\left(g_{m',k}(\gamma_{m',k}, \lambda) \leq g_{m,k}(\gamma_{m,k}, \lambda), \forall m' \neq m\right)$. Therefore, using this pdf, we can easily compute the per-user ergodic rate as

$$\bar{R}_m = \sum_{k \in \mathcal{K}} \int_{\gamma_{0,m}}^\infty \log_2 \left(\frac{\gamma_{m,k}}{\gamma_{0,m}} \right) f_{\gamma_{m,k}}(\gamma_{m,k}) \prod_{m' \neq m} F_{\gamma_{m',k}}(\check{\gamma}_{m',k}(g_{m,k}(\gamma_{m,k}, \lambda))) d\gamma_{m,k}$$

$$(5.20)$$

where we omitted the $[\cdot]^+$ operation since the integration is performed over $\gamma_{m,k} > \gamma_{0,m}$. If desired, the average power per-user can also be computed as

$$\bar{P}_m = \sum_{k \in \mathcal{K}} \int_{\gamma_{0,m}}^\infty \left(\frac{1}{\gamma_{0,m}} - \frac{1}{\gamma_{m,k}} \right) f_{\gamma_{m,k}}(\gamma_{m,k}) \prod_{m' \neq m} F_{\gamma_{m',k}}(\check{\gamma}_{m',k}(g_{m,k}(\gamma_{m,k}, \lambda))) d\gamma_{m,k}$$

$$(5.21)$$

5.2.4 Complexity Analysis

The complexity analysis proceeds similarly as in Sec. 3.2.5, with an additional outer iteration of the subgradient search for $\boldsymbol{\mu}^*$. Furthermore, each outer iteration requires the computation of M per-user ergodic rates, which when assuming NIID channel gains across subcarriers and N function evaluations per integral with $\mathcal{O}(M)$ operations, has $\mathcal{O}(NM^2)$ complexity. Thus, letting I_μ denote the number of subgradient search iterations to find $\boldsymbol{\mu}^*$ and I_λ the number of line-search iterations to find λ^*, the overall initialization complexity is $I_\mu(I_\lambda\mathcal{O}(NM) + \mathcal{O}(NM^2))$, which has order $\mathcal{O}(I_\mu NM^2)$. The per-symbol processing complexity is identical to that of the weighted sum-rate case, which is $\mathcal{O}(MK)$, since we simply replace the weights w_m by μ_m^*. Thus, considering proportional rates only increases the initialization complexity, since we have to find the optimal multipliers $\boldsymbol{\mu}^*$.

5.2.5 Extension to Discrete Rates and/or Imperfect CSI

We have shown that considering proportional rates is essentially a weighted-sum rate problem with the optimal weights given as the dual optimal multipliers $\boldsymbol{\mu}^*$. Hence, using the same subgradient search technique, the extension to the discrete rate case similar to Sec. 3.3 and the extension to assuming imperfect CSI similar to Chap. 4 can be easily performed.

5.3 Adaptive algorithms for Rate Maximization without CDI

In the previous section (and in Chapters 3-4), we assumed the availability of the channel distribution information (CDI) at the transmitter. Although there are methods that allow us to estimate this (e.g. goodness-of-fit tests followed by maximum likelihood parameter estimation [93]), they are typically quite computationally intensive, and are more suitable for offline processing. In our scenario, it is important to be able to perform the resource allocation in real-time, hence online adaptive algorithms are more desirable. In this section, we outline a framework based on *stochastic approximation* to perform adaptive OFDMA resource allocation that allows us to do without the CDI. Note that stochastic approximation methods have been studied in the context of wireless network scheduling for TDMA in [59], and for weighted-sum continuous rate maximization for a downlink OFDMA system [44].

5.3.1 Overview of Stochastic Approximation

Stochastic approximation methods (see e.g. [94]) have been studied extensively since the first algorithms introduced by Robbins and Monro in the early

1950s [96]. The fundamental principle behind these algorithms is a stochastic difference equation of the form

$$\theta[n+1] = \theta[n] + \epsilon[n]Y[n] \tag{5.22}$$

where $\theta[n]$ is some real parameter, $Y[n]$ is some observation random variable, and $\epsilon[n] > 0$ is some small step size that may be diminishing to zero. Under some mild conditions, it can be shown that the iterates converge to some stationary point $\lim_{n\to\infty} \theta[n+1] = \theta^*$ w.p.1. In our resource allocation algorithms, we are interested in finding recursions for the multipliers λ and μ which in the limit converge to the optimal values that solve the dual problem. This is the goal of the subsequent section.

5.3.2 Stochastic Approximation Solution to the Dual Problem

The purpose of this section is to derive suitable stochastic approximation recursions that solve the dual problem given in (5.7) without knowledge of the pdf of γ. The objective is to construct a sequence of approximants $\lambda[n]$ and $\mu[n]$, $n = 0, 1, \ldots$ using statistic estimates of the subgradients $g_\lambda[n]$ in (5.10) and $g_\mu[n]$ in (5.12).

The fundamental stochastic approximation iteration we employ is based on the subgradient iterations given in (5.8) and (5.9), but performed across time, i.e.

$$\lambda[n+1] = [\lambda[n] - \beta_n g_\lambda[n]]_\epsilon^+ \tag{5.23}$$

$$\mu[n+1] = \Pi_{\mathcal{U}} [\mu[n] - \beta_n g_\mu[n]] \tag{5.24}$$

where $[x]_\epsilon^+ = \max(x, \epsilon)$ for a small constant $0 < \epsilon \ll 1$ and is used in (5.23) as a modified projection operator to prevent λ from going to zero (which results in infinite power), and β_n is a real-valued step-size chosen to satisfy

$$\sum_{n=0}^{\infty} \beta_n = \infty, \beta_n \geq 0, \beta_n \to 0 \tag{5.25}$$

Furthermore, we employ an auxiliary filter to perform *subgradient averaging*

$$\begin{aligned} g_\lambda[n+1] &= (1 - \alpha_n)g_\lambda[n] + \alpha_n \hat{g}_\lambda[n] \\ &= g_\lambda[n] + \alpha_n(\hat{g}_\lambda[n] - g_\lambda[n]) \end{aligned} \tag{5.26}$$

$$\begin{aligned} g_\mu[n+1] &= (1 - \alpha_n)g_\mu[n] + \alpha_n \hat{g}_\mu[n] \\ &= g_\mu[n] + \alpha_n(\hat{g}_\mu[n] - g_\mu[n]) \end{aligned} \tag{5.27}$$

with α_n as a non-negative step-size chosen to satisfy

$$\alpha_n \geq 0, \frac{\beta_n}{\alpha_n} \to 0, \sum_{n=0}^{\infty}(\beta_n^2 + \alpha_n^2) < \infty \tag{5.28}$$

and where $\hat{g}_\lambda[n]$ and $\hat{\boldsymbol{g}}_\mu[n]$ are approximations to the subgradient given the current CNR realization $\boldsymbol{\gamma}[n]$ and the current estimates for the multipliers $\lambda[n]$ and $\boldsymbol{\mu}[n]$. This method that employs averaging of the search directions are called averaged, aggregated, or mixed stochastic gradient or quasigradient methods [63, Sec. 6.2.4] [97]. Note that the conditions on step sizes α_n and β_n are to ensure w.p.1 convergence, which will be discussed in Sec. 5.3.3. A possible choice is given by

$$\beta_n = \frac{b_1}{b_2 + n} \tag{5.29}$$

$$\alpha_n = \frac{a_1}{a_2 + n^{0.4}} \tag{5.30}$$

with real constants $a_1 > 0$, $a_2 \geq 0$, $b_1 > 0$, and $b_2 \geq 0$.

A suitable approximation to the subgradient would be to replace the expectations with the instantaneous (sample) subgradient, which can be computed via a single iteration of the "multi-level waterfilling" with "max-dual user selection" (3.14)-(3.15) procedure. We repeat this operation here for convenience:

$$\tilde{p}_{m,k}[n] = \left[\frac{\mu_m[n]}{\lambda[n] \ln 2} - \frac{1}{\gamma_{m,k}[n]} \right]^+ \tag{5.31}$$

$$m_k^*[n] = \arg \max_{m \in \mathcal{M}} \left\{ \mu_m[n] R_{m,k} \left(\tilde{p}_{m,k}[n] \gamma_{m,k}[n] \right) - \lambda[n] \tilde{p}_{m,k}[n] \right\} \tag{5.32}$$

$$p_{m,k}^*[n] = \begin{cases} \tilde{p}_{m,k}[n], & m = m_k^*[n] \\ 0, & \text{otherwise} \end{cases} \tag{5.33}$$

where we use $\gamma_{m,k}[n]$ to denote the channel gain for user m and subcarrier k at time n. Observe that in the process of our stochastic subgradient iterations, we also generate the resource allocation procedure for time n given by (5.31)-(5.33).

The per-user instantaneous rate is then given as

$$R_m[n] = \sum_{k \in \mathcal{K}} R_{m,k} \left(p_{m,k}^*[n] \gamma_{m,k}[n] \right) \tag{5.34}$$

with instantaneous total power

$$P[n] = \sum_{m \in \mathcal{M}} \sum_{k \in \mathcal{K}} p_{m,k}^*[n] \tag{5.35}$$

The subgradient approximations are then given as

$$\hat{g}_\lambda[n] = \bar{P} - P[n] \tag{5.36}$$

$$\hat{\boldsymbol{g}}_\mu[n] = \boldsymbol{R}[n] - \phi R[n] \tag{5.37}$$

where $\boldsymbol{R}[n] = [R_1[n], \ldots, R_M[n]]^T$ and $R[n] = \sum_{m \in \mathcal{M}} R_m[n]$. Using (5.36)-(5.37) in the subgradient averaging operations (5.26)-(5.27) completes our algorithm. Fig. 5.1 shows the block diagram for the proposed algorithm.

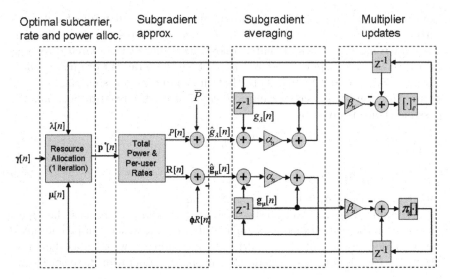

Fig. 5.1. Block diagram for adaptive OFDMA resource allocation for ergodic sumrate maximization with ergodic proportional rate constraints.

It is interesting to note that this stochastic approximation procedure can similarly be applied to the simpler weighted sum-rate formulations in Ch. 3-4 by using the update procedure on λ (5.23) and using the appropriate power, subcarrier, and rate (if applicable) allocation procedures per iteration given the current $\lambda[n]$.

5.3.3 Proof of Convergence

The convergence proof for this stochastic approximation procedure under various assumptions is quite well studied [63] [94] [98]. We repeat one such convergence theorem from [63, Sec. 6.2.4] as our basis.

Theorem 5.1. *Suppose we have to minimize a convex continuous function* $f(x)$ *such that* $x \in \mathcal{X} \subset \mathbb{R}^n$, *where* $\mathcal{X}$ *is a closed convex set such that the projection on* $\mathcal{X}$ *can easily be calculated:* $\Pi_{\mathcal{X}}[y] = \arg\min\left[\|y - x\|^2 | x \in \mathcal{X}\right]$. *Let* $\mathcal{X}^*$ *be the set of optimal solutions. Consider the relations*

$$x[n + 1] = \Pi_{\mathcal{X}}\left[x[n] - \beta_n g_x[n]\right] \tag{5.38}$$

$$g_x[n + 1] = (1 - \alpha_n)g_x[n] + \alpha_n \hat{g}_x[n] \tag{5.39}$$

$$\mathbb{E}_{\gamma[n]}\left\{\hat{g}_x[n] \middle| x[0], g_x[0], \dots, x[n], g_x[0]\right\} = g_x(x[n]) + e[n] \tag{5.40}$$

where $g_x(x[n])$ *is a subgradient of* f *with respect to* x *evaluated at* $x[n]$ *and* $e[n]$ *is some random variable that can be interpreted as the subgradient approximation error, and* β_n *and* α_n *satisfy (5.25) and (5.28), respectively. If in addition, we assume*

$$\sum_{n=0}^{\infty} \mathbb{E}\left\{\beta_n |\langle e[n], \boldsymbol{x}^* - \boldsymbol{x}[n]\rangle| + \beta_n^2 \|\hat{\boldsymbol{g}}_{\boldsymbol{x}}[n]\|^2\right\} < \infty \tag{5.41}$$

then $x[n] \to \boldsymbol{x}^*$ w.p.1.

The dual objective $\Theta(\lambda, \boldsymbol{\mu})$ is convex and continuous, and $\lambda \in \mathbb{R}^+$ and $\boldsymbol{\mu} \in \mathcal{U}$ are subspace constraints with simple projections, therefore, what is left to show is that the subgradient approximations fulfill (5.40)-(5.41). Expanding the left-hand side of (5.40) for the subgradient with respect to λ, we have

$$\begin{aligned}
\mathbb{E}_{\boldsymbol{\gamma}[n]}\left\{\hat{g}_\lambda[n]|\,\lambda[0], \boldsymbol{\mu}[0], \ldots\right\} &= \mathbb{E}_{\boldsymbol{\gamma}[n]}\left\{\bar{P} - P[n]|\lambda[n], \boldsymbol{\mu}[n]\right\} \\
&= \bar{P} - \mathbb{E}_{\boldsymbol{\gamma}}\left\{P[n]|\lambda[n], \boldsymbol{\mu}[n]\right\} \\
&= g_\lambda(\lambda[n], \boldsymbol{\mu}[n])
\end{aligned} \tag{5.42}$$

where the second equality is due to the stationarity of $\boldsymbol{\gamma}[n]$, and the third equality is due to (5.10)-(5.11). The subgradient with respect to $\boldsymbol{\mu}$ likewise follows (see (5.12)-(5.14)):

$$\begin{aligned}
\mathbb{E}_{\boldsymbol{\gamma}[n]}\left\{\hat{g}_{\boldsymbol{\mu}}[n]|\,\lambda[0], \boldsymbol{\mu}[0], \ldots\right\} &= \mathbb{E}_{\boldsymbol{\gamma}[n]}\left\{\boldsymbol{R}[n] - \phi R[n]|\lambda[n], \boldsymbol{\mu}[n]\right\} \\
&= g_{\boldsymbol{\mu}}(\lambda[n], \boldsymbol{\mu}[n])
\end{aligned} \tag{5.43}$$

The subgradient approximation errors are zero for both cases, and thus our method belongs to a class of stochastic approximation algorithms called *stochastic subgradient averaging methods*. Finally, we need to show that (5.41) holds for both subgradient approximants. For the subgradient with respect to λ, we have

$$\begin{aligned}
\sum_{n=0}^{\infty} \mathbb{E}\left\{\beta_n^2 |\hat{g}_\lambda[n]|^2\right\} &= \sum_{n=0}^{\infty} \beta_n^2 \mathbb{E}\left\{(\bar{P} - P[n])^2\right\} \\
&= \sum_{n=0}^{\infty} \beta_n^2 \left(\bar{P}^2 - \bar{P}\mathbb{E}\left\{P[n]\right\} + \mathbb{E}\left\{P[n]^2\right\}\right) \\
&\leq \bar{P}^2 \sum_{n=0}^{\infty} \beta_n^2 + \sum_{n=0}^{\infty} \beta_n^2 \mathbb{E}\left\{P[n]^2\right\}
\end{aligned} \tag{5.44}$$

The first term is clearly bounded by our choice of step-size in (5.28), and

$$\begin{aligned}
\mathbb{E}\left\{P[n]^2\right\} &= \mathbb{E}\left\{\left(\sum_{m \in \mathcal{M}} \sum_{k \in \mathcal{K}} p^*_{m,k}[n]\right)^2\right\} \\
&\leq (MK)^2 \mathbb{E}\left\{\left(\max_{m,k} \tilde{p}_{m,k}[n]\right)^2\right\} \\
&\leq (MK)^2 \left(\max_m \frac{\mu_m[n]}{\ln 2\lambda[n]}\right)^2
\end{aligned} \tag{5.45}$$

is also bounded since $\mu_m[n] \le B_\mu < \infty, \forall n$ and we have $\lambda[n] \ge \epsilon, \forall n$ by our update given in (5.23). Therefore, we have

$$\sum_{n=0}^{\infty} \mathbb{E}\left\{\beta_n^2 |\hat{g}_\lambda[n]|^2\right\} \le \bar{P}^2 \sum_{n=0}^{\infty} \beta_n^2 + \sum_{n=0}^{\infty} \beta_n^2 \mathbb{E}\left\{P[n]^2\right\}$$

$$\le \bar{P}^2 \sum_{n=0}^{\infty} \beta_n^2 + \left(MK\frac{B_\mu}{\epsilon \ln 2}\right)^2 \sum_{n=0}^{\infty} \beta_n^2 \qquad (5.46)$$

$$< \infty$$

We can similarly bound the subgradient with respect to $\boldsymbol{\mu}$ using a similar approach, and is skipped in the interest of brevity. Therefore, we arrive at the following proposition:

Proposition 5.2. *Consider the dual problem* (5.7). *The iterations for λ and $\boldsymbol{\mu}$ given in* (5.23)-(5.24), *with stochastic subgradient averaging given in* (5.26)-(5.27), *and step-size criteria* (5.25) *and* (5.28), *converges w.p.1 to the unique optimal values λ^* and $\boldsymbol{\mu}^*$.*

The convergence proof we presented here actually uses some restrictive assumptions, including the ergodicity and stationarity of the CNRs. Furthermore, the step-size ratio requirement of $\beta_n/\alpha_n \to 0$ causes a degradation of the local rate of convergence (since the rate is determined by the slowest part of the algorithm) [98], and the decreasing step-size requirements makes it more difficult to use the algorithm as a tracking mechanism when encountering non-stationary CNR statistics [94]. Fortunately, the most recent convergence results given in [94] actually allows more relaxed assumptions, including the use of small constant step-sizes to improve tracking capability. We thus use the constant step-size rules in the simulations.

5.3.4 Complexity Analysis

The complexity of this algorithm is significantly lower than our algorithm assuming perfect CDI, since all that is needed is the multi-level waterfilling and max-dual user selection with $\mathcal{O}(MK)$, followed by $\mathcal{O}(M)$ updates for the rates, power, and multipliers. Hence, we do away completely with the initialization complexity, and have allowed our "iterations" to be performed over time and on the fly. This holds true also with the weighted-sum rate problems in Ch. 3-4.

5.3.5 Extension to Other Formulations

Although we developed in detail the adaptive algorithm for continuous sum-rate maximization with proportional rate constraints, it is relatively straightforward to extend the algorithm to the discrete rate and/or partial CSI proportional rate and weighted-sum rate cases. The required changes are: (1) using

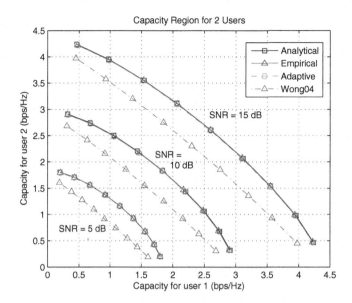

Fig. 5.2. Two-user capacity region for ergodic sum-rate maximization with proportional rate constraints.

the appropriate optimal resource allocation algorithms developed in Chapters 3-4 for computing the approximate subgradients; (2) for the weighted-sum rate maximization cases, drop the subgradient iterations for μ, and use the λ subgradient updates instead of the line-search procedures to find λ^*. When using the adaptive algorithm in these problems, the complexity is $\mathcal{O}(MK)$ per OFDMA symbol without the need for iterations, and are thus the lowest complexity algorithms available for asymptotically optimal resource allocation for OFDMA systems.

5.4 Results and Discussion

We use the same simulation assumptions as in Sec. 4.5. Fig. 5.2 shows the $M = 2$ user capacity region with $\phi_1 = 0.1$ to $\phi_1 = 0.9$ in 0.1 increments and $\phi_2 = 1 - \phi_1$ for the following:

1. *Analytical*: Numerical evaluation of the per-user ergodic rate integral (5.20)
2. *Empirical*: Sample average of the per-user rates by using the pre-computed λ^* and μ^*
3. *Adaptive*: Sample average of the per-user rates of the algorithm in Sec. 5.3.2 with constant step-size $\alpha_n = \beta_n = 0.005$
4. *Wong04*: Sample average of the per-user rates using the current state-of-the-art algorithm for proportional rate OFDMA resource allocation [46]

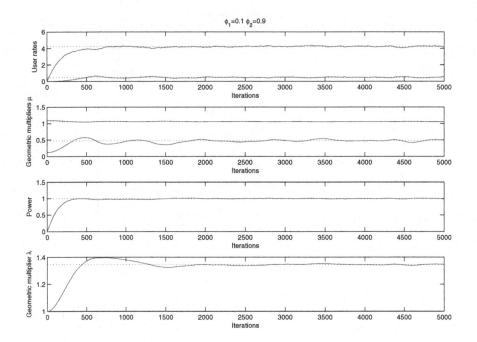

Fig. 5.3. Evolution across iterations of the exponentially averaged user rates and power, and their corresponding geometric multipliers. Theoretical values solved using a perfect CDI assumption is shown in dotted lines.

Observe that in contrast to the weighted-sum rate capacity regions (Fig. 3.12), the rate points for all the methods are neatly spaced along the boundary of the rate region since we constrain $\bar{R}_1/\bar{R}_2 = \phi_1/\phi_2$, confirming that the algorithms indeed enforce the proportional rate constraints. We also observe that methods 1-3 give essentially identical results, confirming our analysis in the previous sections. On the other hand, using a per-symbol algorithm [46], which is more complex than our algorithms, has significantly poorer performance, because it is suboptimal to start with, and that it is unable to exploit the temporal dimension.

Fig. 5.3 shows the evolution of the exponentially averaged user rates $\bar{R}_m[n] = (1-\beta_n)\bar{R}_m[n-1]+\beta_n R_m[n]$ and average power $\bar{P}[n] = (1-\beta_n)\bar{P}[n-1] + \beta_n P[n]$, together with the multipliers $\lambda[n]$ and $\boldsymbol{\mu}[n]$ with initializations $\lambda[0] = \bar{P}$, $g_\lambda[0] = 0$, $\boldsymbol{\mu}[0] = \boldsymbol{\phi}/(\boldsymbol{\phi}^T\boldsymbol{\phi})$, and $\boldsymbol{g}_\mu[0] = \mathbf{0}$ for an SNR of 15 dB and for proportionality constants $\boldsymbol{\phi} = [0.1, 0.9]^T$ (the results are similar for other $\boldsymbol{\phi}$ values). We can see that the iterates converge to their offline-equivalent optimal values, which are shown by the dotted lines.

5.5 Conclusion

In this chapter, we proposed optimal algorithms to maximize the ergodic sum rates subject to proportional rate constraints. The important contributions of this chapter are:

- *Optimal continuous sum-rate maximization with proportional continuous rate constraints assuming perfect CSI and CDI:* We derived the optimal algorithm for OFDMA resource allocation for ergodic sum-rate maximization subject to ergodic rate proportionality constraints. It is shown that the proportional rates can be enforced by a weighted-sum rate formulation using optimally chosen weights, which are themselves the dual-optimal geometric multipliers. Therefore, we can use similar algorithms developed in Chapter 3, in addition to a subgradient search technique to determine the optimal weights.
- *Extensions to proportional discrete rate cases and extension to assuming partial CSI:* Since the proportional rate constraints can be enforced using a weighted-sum rate formulation, the extensions to discrete rates and/or partial CSI cases can be performed using the techniques developed in Section 3.3 and Chapter 4.
- *Adaptive resource allocation for proportional rate constraints assuming perfect CSI but without CDI:* We developed an adaptive algorithm that updates the geometric multipliers over time using a subgradient search and stochastic subgradient averaging. It is based on general *stochastic approximation* principles, and is shown to converge to the optimal solution w.p.1.
- *Adaptive resource allocation for weighted-sum rate formulations:* The adaptive algorithm developed is shown to be general enough to encompass all of the previous formulations considered by using the subgradient iterations across time for the geometric multiplier that enforces the average power constraint. Thus, we have developed a truly adaptive resource allocation algorithm that requires only linear complexity per symbol *without iterations*, and can be considered the algorithms with the lowest complexity while assuring asymptotic optimality in OFDMA.

6

Conclusion

6.1 Summary

In this book, we proposed a common framework for resource allocation in M-user and K-subcarrier OFDMA systems with $\mathcal{O}(MK)$ complexity that achieve negligible optimality gaps in simulations based on realistic parameters. The main assumptions of the book are:

1. Stationarity and ergodicity of channel gains
2. Statistical independence of channel gains across users
3. Absence of inter-cell interference
4. MAC layer provides the active user set and the user priorities in the form of weights or proportionality constants

The framework is based on dual optimization techniques, and was shown to apply to a wide variety of OFDMA resource allocation problem formulations, including:

1. Ergodic/instantaneous weighted-sum continuous/discrete rate maximization with perfect CSI and CDI (Ch. 3)
2. Ergodic/instantaneous weighted-sum continuous/discrete rate maximization with partial CSI and CDI (Ch. 4)
3. Sum-rate maximization with proportional rate constraints with or without CDI (Ch. 5)

Table 6.1 repeats Table 2.1 in Sec. 2.2 which compares the related literature, but this time together with the proposed algorithms in this book. Observe that we are able to improve on the state-of-the-art by considering formulations with ergodic and discrete rates, by considering solutions that are of linear complexity yet achieve negligible optimality gaps in simulation, and by considering partial CSI cases even in the absence of CDI.

The primary reason dual methods work well in OFDMA problems is due to the problem structure, i.e. there are typically a lot more subcarriers K than users M. In most OFDMA/multicarrier resource allocation problems,

Table 6.1. Comparison of proposed algorithms with previous work

Criteria / Method	Formulation			Solution		Assumption	
	(1)	(2)	(3)	(4)	(5)	(6)	(7)
Max-min rate[41]	No	No	No	No	No	No	Yes
Sum rate [37][38]	Yes	No	No	Yes	No	No	No
Proportional rate[45][46]	No	No	Yes	No	No	No	Yes
Max-utility [54][55]	[a]No	Yes	Yes	No	No	No	Yes
Weighted rate[43][44]	No	No	Yes	[b]Yes	[c]Yes	No	Yes
Perf. CSI, Cont. Rate (Sec. 3.2)	Yes	No	Yes	Yes	Yes	No	No
Perf. CSI, Disc. Rate (Sec. 3.3)	Yes	Yes	Yes	Yes	Yes	No	No
Imperf. CSI, Cont. Rate (Sec. 4.3)	No	No	Yes	Yes	Yes	Yes	No
Imperf. CSI, Disc. Rate (Sec. 4.4)	No	Yes	Yes	Yes	Yes	Yes	No
Adaptive, Imperf. CSI Disc. Rate (Sec. 5.3)	Yes	Yes	Yes	Yes	Yes	Yes	Yes

[a] Considered some form of temporal diversity by maximizing an exponentially windowed running average of the rate
[b] Independently developed a similar instantaneous continuous rate maximization algorithm
[c] Only for instantaneous continuous rate case, but was not shown in their papers

Criteria

(1) Ergodic rates: The optimization problem is posed such that the *expected value* of the rate is being maximized instead of *instantaneous rate*, which allows the temporal dimension to be exploited when assuming ergodicity of channel gains.
(2) Discrete rates: The practical transmission scheme of only allowing a discrete set of possible data rates is considered rather than just the theoretical continuous rate.
(3) User prioritization: The problem formulation allows setting varying priorities among users to ensure fairness in the system.
(4) Practically optimal: The algorithm is shown in simulations using realistic parameters to have negligible optimality gaps.
(5) Linear complexity: The algorithm can be performed with complexity that is just linear in the number of users and subcarriers.
(6) Imperfect CSI: The algorithm assumes the more realistic scenario of the presence of errors in the available channel state information.
(7) Does not require CDI: The algorithm does not assume knowledge of the probability distribution function of the channel gains, which is difficult to obtain in practice.

the objective function is separable across the K subcarriers, and the number of constraints are in the order of the number of users M. This makes dual optimization techniques an ideal approach to solving them, since the duality gap is typically quite small in these types of problems, as shown in Sec. 3.2.8. Furthermore, we also saw that the solution to the dual problem involves very simple closed-form power, subcarrier, and rate allocation functions for both continuous and discrete rates, thus further enhancing the attractiveness of using a dual approach.

Although the dual approach is very useful and widely applicable, there are some useful problem formulations in OFDMA wherein the separability of the objective function across subcarriers do not hold, and thus limits the applicability of the dual approach. An example of this is the maximization of the *utility function* of the rates, wherein the utility function is something other than linear (see e.g. [55]). Fortunately, using the stochastic approximation methods for adaptive resource allocation, it has been shown in [55] that as long as the chosen step size is small, a first order Taylor expansion of the utility function results in an equivalent weighted-sum ergodic rate formulation, thus our proposed adaptive methods can also be used in this context.

6.2 Future Work

In this chapter, we outline several interesting research directions that this book can be extended to.

6.2.1 Other Formulations

Uplink OFDMA

In uplink OFDMA, the single average power constraint is replaced with per-user power constraints. In this case, the separability of the objective function across the subcarriers is still possible. This is done by using a vector of geometric multipliers $\boldsymbol{\lambda} = [\lambda_1, \ldots, \lambda_M]^T$, where each multiplier enforces the per-user power constraint $\sum_{k \in \mathcal{K}} p_{m,k} \leq \bar{P}_m, \forall m \in \mathcal{M}$, resulting in the new dual objective

$$\Theta(\boldsymbol{\lambda}) = \sum_{m \in \mathcal{M}} \lambda_m \bar{P}_m + \sum_{k \in \mathcal{K}} \max_{m \in \mathcal{M}} \max_{p_{m,k} \geq 0} \mathbb{E}_{\gamma_{m,k}} \{ w_m R_{m,k} (p_{m,k} \gamma_{m,k}) \} - \lambda_m p_{m,k}$$

(6.1)

This is essentially identical to our downlink dual problems, and have the same dual optimal solutions. The main difference is that we are now searching for a *vector* of geometric multipliers, instead of just a single one, similar to the case in Ch. 5. Thus, we can use the subgradient search technique developed in Sec. 5.2.2, and the resulting asymptotic complexity is still $\mathcal{O}(MK)$ per iteration, although it will take longer to attain convergence.

Non-real-time traffic

In non-real-time traffic, e.g. file transfers, minimum rate constraints typically need to be enforced [59]. Suppose we also wish to maximize a weighted-sum rate, and assuming that appropriate admission control is performed such that the minimum rates are feasible (i.e. it is within the capacity region), then a similar vector of multipliers $\boldsymbol{\rho} = [\rho_1, \ldots, \rho_M]^T$ can be used to enforce the minimum rate constraints, resulting in the dual objective

$$\Theta(\lambda, \boldsymbol{\rho}) = \lambda \bar{P} + \sum_{k \in \mathcal{K}} \max_{m \in \mathcal{M}} \max_{p_{m,k} \geq 0} \mathbb{E}_{\gamma_{m,k}} \left\{ (w_m + \rho_m) R_{m,k} \left(p_{m,k} \gamma_{m,k} \right) \right\} - \lambda p_{m,k}$$

(6.2)

which is essentially in the form of the dual objective for our proportional rates case (5.6), except for the fact that the equivalent "user-weight" is now $w_m + \rho_m$. The optimal $\boldsymbol{\rho}^*$ can also be searched using the subgradient technique.

In these aforementioned formulations, continuous or discrete rates, perfect or predicted CSI, perfect or no CDI cases are all readily available given our developed algorithms in this book. One caveat, though, is that the duality gaps in these cases will be higher, since the gaps scale with the number of dualized constraints as shown in Sec. 3.2.8. However, as long as $K \gg M$, the solutions should still be near-optimal.

Power or BER Minimization

Although this book focused on the capacity maximization problem, the developed framework is extendable to power/BER minimization. Since the average power and average BER are similarly separable objectives across the subcarriers, a similar dual optimization approach can be used to find the power, subcarrier, and/or rate allocations. Note that the instantaneous minimum weighted-sum power problem in OFDMA is solved using dual methods in [43].

Non-homogenous traffic types and services

Another interesting venue of future investigation is when various traffic types and services all compete for the limited amount of resources. In these scenarios, multiple objectives need to be met, e.g. maximizing the throughput, minimizing the delay, or minimizing the transmit power. Although Pareto optimality is the desirable optimality criterion in multi-objective optimization, simplified formulations that involve maximizing the sum of several utility functions have been shown to perform well in practice for the TDMA/FDMA cases in [59]. The extension to the OFDMA scenario will be an interesting avenue for further investigation.

Outage Capacity

This book focused on the ergodic capacity maximization problem. In some cases, we would like to maximize the outage capacity instead. These problems involve probabilistic constraints instead of the ergodic constraints, and are typically harder to solve. It is interesting to study if the dual optimization approach can still be used in these types of problems.

6.2.2 MAC-PHY Cross-layer Scheduling

We have focused primarily on the physical layer transmit optimization problem, and simply assumed that the upper MAC layer is responsible for performing admission and congestion control (the number of active users M is given to us), and user prioritization (by setting the user weights or proportionality constants). We have provided the necessary tools in the physical layer for transmit optimization, but it is interesting to see the overall performance at the network level for various traffic types and services, and including the effect of finite queue lengths.

6.2.3 MIMO-OFDMA

This book has focused on OFDMA systems with single transmit and receive antennas. Extending this to systems with multiple transmit and receive antennas, i.e. MIMO-OFDMA [99], is certainly an interesting problem.

The algorithms proposed in this book can be extended in straightforward manner when transmit beamforming with maximal ratio combining is used, since all this changes is the pdf of the CNR per subcarrier. In the case of using "spatial waterfilling", however, requires a slight extension of the proposed algorithms. In narrowband N_t-transmit and N_r-receive antenna MIMO transmission assuming perfect CSI, it is well-known that the optimal transmit covariance matrix is the sum of appropriately weighted outer products of the right singular vectors of the $N_r \times N_t$ channel matrix [99], where the weights are found using spatial waterfilling across the channel eigenmodes. In MIMO-OFDMA, we can model the frequency-selective channel by a separate $N_r \times N_t$ channel matrix per subcarrier. By performing an singular value decomposition (SVD) per-subcarrier, we would then have $\min(N_r, N_t)$ (assuming a full-rank channel matrix) "spatial gains" per subcarrier. Thus, instead of the K CNR values in the SISO-OFDMA case, we have $K \min(N_r, N_t)$ CNR values, and we can reuse the dual optimization methods developed for the SISO-OFDMA case, since we can similarly use "multi-level waterfilling" power allocation across all spatial gains of all subcarriers, and similarly assign the subcarrier to the user that maximizes the marginal dual, which in this case is computed as the MIMO capacity per-subcarrier. Using other MIMO transmission methods, e.g. space-time coding and spatial multiplexing, can also be solved using the dual optimization framework, as long as the objective function is separable across the subcarriers. Developing efficient algorithms for these cases are interesting avenues for further research.

6.2.4 Multi-cell OFDMA

In this book, we focused on the single-cell allocation scenario, and ignored the effect of inter-cell interference. In practical networks, inter-cell interference significantly affect the performance, and thus should be properly controlled, avoided, and/or canceled. Although centralized control may not be feasible from a practical perspective, a theoretical study of the capacity region assuming that all resource allocation across cells can be performed centrally is interesting. In this case, the dual optimization framework is still applicable, however, the power allocation procedure would involve a non-convex per-user and per-subcarrier non-linear program for the continuous rate case, or exhaustive search of possible bit allocations in the discrete rate case.

A more interesting research study would be in the semi-coordinated case, wherein a small amount of information, e.g. the loading of the cell or the total interference power experienced by a neighboring cell, is exchanged across base stations. In this case, we can use the algorithms proposed in this book, with the addition of a "penalty" term that is a function of the information that is exchanged, such that the marginal dual is decreased for problematic users, e.g. users that are in the cell-edge. Game-theoretic algorithms and analysis and low complexity adaptive algorithms using stochastic approximation can be effectively used in these problems.

6.2.5 Multi-hop OFDMA

An interesting method to increase the coverage of cellular networks is the concept of *multi-hop radio*, wherein relay nodes are scattered around strategic areas in a cell, and are used to improve the signal reliability of a subscriber node by "relaying" the data to/from the base-station. In these multi-hop networks, an additional degree of freedom, namely, link selection, is introduced as a means of improving the overall system capacity. From a resource allocation perspective, this means that for each subcarrier, in addition to deciding which user can transmit, we also need to decide which link to transmit over. There are also several important system design issues, e.g. what type of frequency reuse method and what type of relaying to use, which makes for interesting future investigations.

A

Derivation of the inverse function (3.10) of $g_{m,k}$ (3.8)

Since $g_{m,k}$ for $\gamma_{m,k} \geq \gamma_{0,m}$ is monotonically increasing and non-negative, there exists a unique inverse function. After some algebraic manipulation, we have

$$-\frac{\gamma_{0,m}(\lambda)}{\gamma_{m,k}} e^{\left(-\frac{\gamma_{0,m}(\lambda)}{\gamma_{m,k}}\right)} = -e^{\left(-g_{m,k}\frac{\ln 2}{w_m}-1\right)} \tag{A.1}$$

Observe that this is in the form of the Lambert-W function $W(x)$ [68], which is the solution to $W(x)e^{(W(x))} = x$. Thus, we can write

$$W\left(-e^{\left(-g_{m,k}\frac{\ln 2}{w_m}-1\right)}\right) = -\frac{\gamma_{0,m}(\lambda)}{\gamma_{m,k}} \tag{A.2}$$

which when solved for $\gamma_{m,k}$ gives us (3.10).

B

Proof of Prop. 3.1

From the assumption of constraint tightness, we have

$$\bar{P} = \sum_{k \in \mathcal{K}} \left[\frac{w_{m_k^*}}{\lambda_{\text{inst}}^* \ln 2} - \frac{1}{\gamma_{m_k^*, k}} \right]^+$$

$$\geq \sum_{k \in \mathcal{K}} \min_m \left[\frac{w_m}{\lambda_{\text{inst}}^* \ln 2} - \frac{1}{\gamma_{m,k}} \right]^+$$

$$\geq \sum_{k \in \mathcal{K}} \left(\frac{\min_m w_m}{\lambda_{\text{inst}}^* \ln 2} - \max_m \frac{1}{\gamma_{m,k}} \right) \qquad \text{(B.1)}$$

$$= K \frac{\min_m w_m}{\lambda_{\text{inst}}^* \ln 2} - \sum_{k \in \mathcal{K}} \max_m \frac{1}{\gamma_{m,k}}$$

$$\lambda_{\text{inst}}^* \geq \frac{\ln 2}{K \min_m w_m} \left(\bar{P} + \sum_{k \in \mathcal{K}} \max_m \frac{1}{\gamma_{m,k}} \right)$$

giving us the left inequality in (3.20). Also,

$$\bar{P} = \sum_{k \in \mathcal{K}} \left[\frac{w_{m_k^*}}{\lambda_{\text{inst}}^* \ln 2} - \frac{1}{\gamma_{m_k^*, k}} \right]^+$$

$$\leq \sum_{k \in \mathcal{K}} \max_m \left[\frac{w_m}{\lambda_{\text{inst}}^* \ln 2} - \frac{1}{\gamma_{m,k}} \right]^+ \qquad \text{(B.2)}$$

$$\leq \frac{K}{\lambda_{\text{inst}}^* \ln 2} \max_m w_m$$

$$\lambda_{\text{inst}}^* \leq \frac{K}{\bar{P} \ln 2} \max_m w_m$$

giving us the right inequality in (3.20).

C

Derivation of (3.35)

Observe that (4.36) implies

$$w_m r_{l^*_{m,k}} - \frac{\lambda \eta_{l^*_{m,k}}}{\gamma_{m,k}} \geq w_m r_l - \frac{\lambda \eta_l}{\gamma_{m,k}}, \quad \forall l \in \mathcal{L} \setminus l^*_{m,k} \tag{C.1}$$

which after some algebraic manipulation can also be written as

$$\frac{r_{\bar{l}} - r_{l^*_{m,k}}}{\eta_{\bar{l}} - \eta_{l^*_{m,k}}} \leq \frac{\lambda}{w_m \gamma_{m,k}} < \frac{r_{l^*_{m,k}} - r_{\underline{l}}}{\eta_{l^*_{m,k}} - \eta_{\underline{l}}}, \quad \forall \bar{l} > l^*_{m,k}, \quad \forall \underline{l} < l^*_{m,k}$$

$$\Leftrightarrow \max_{l > l^*_{m,k}} \frac{r_l - r_{l^*_{m,k}}}{\eta_l - \eta_{l^*_{m,k}}} \leq \frac{\lambda}{w_m \gamma_{m,k}} < \min_{l < l^*_{m,k}} \frac{r_{l^*_{m,k}} - r_l}{\eta_{l^*_{m,k}} - \eta_l} \tag{C.2}$$

Since the slope $\Delta r / \Delta \eta$ is non-increasing for a concave function, we arrive at (3.35).

D

Derivation of the cdf (3.39) and pdf (3.40) of $g^d_{m,k}$ (3.38)

First, we use (3.35) to get

$$P(l^*_{m,k} = l) = P\left(\frac{r_{l+1} - r_l}{\eta_{l+1} - \eta_l} \leq \frac{\lambda}{w_m \gamma_{m,k}} < \frac{r_l - r_{l-1}}{\eta_l - \eta_{l-1}}\right) \tag{D.1}$$
$$= P\left(s_l < \gamma_{m,k} \leq s_{l+1}\right)$$

where $s_l = \frac{\lambda(\eta_l - \eta_{l-1})}{w_m(r_l - r_{l-1})}$. Then, using the law of total probability [69], we have

$$F_{g^d_{m,k}}(g^d_{m,k}) = \sum_{l \in \mathcal{L}} P(l^*_{m,k} = l) P\left(\max_{l' \in \mathcal{L}} w_m r_{l'} - \lambda\frac{\eta_{l'}}{\gamma_{m,k}} \leq g^d_{m,k} \,\middle|\, l^*_{m,k} = l\right)$$
$$= \sum_{l \in \mathcal{L}} P(l^*_{m,k} = l) P\left(w_m r_l - \lambda\frac{\eta_l}{\gamma_{m,k}} \leq g^d_{m,k} \,\middle|\, s_l < \gamma_{m,k} \leq s_{l+1}\right) \tag{D.2}$$

Note that since $\lambda\frac{\eta_l}{\gamma_{m,k}}$ is non-negative, then if $w_m r_l - g^d_{m,k}$ is negative,

$$P\left(w_m r_l - \lambda\frac{\eta_l}{\gamma_{m,k}} \leq g^d_{m,k}\right) = 1.$$

Hence, we can write

$$P\left(w_m r_l - \lambda\frac{\eta_l}{\gamma_{m,k}} \leq g^d_{m,k}\right) = P\left(\gamma_{m,k} \leq \frac{\lambda\eta_l}{[w_m r_l - g^d_{m,k}]^+}\right),$$

where we can safely define $\frac{x}{0} = \infty, \forall x > 0$. However, for $l = 0$, we have $r_l = 0$ and $\eta_l = 0$, and $P\left(w_m r_l - \lambda\frac{\eta_l}{\gamma_{m,k}} \leq g^d_{m,k}\right)$ is always one since $g^d_{m,k} \geq 0$. We can now write (D.2) as

$$F_{g_{m,k}^d}(g_{m,k}^d) = u(g_{m,k}^d)P(l_{m,k}^* = 0) +$$

$$\sum_{l \in \mathcal{L}\backslash 0} P(l_{m,k}^* = l) \frac{P\left(\gamma_{m,k} \leq \frac{\lambda\eta_l}{[w_m r_l - g_{m,k}^d]^+}, s_l \leq \gamma_{m,k} \leq s_{l+1}\right)}{P(l_{m,k}^* = l)}$$

$$= u(g_{m,k}^d)P(\gamma_{m,k} \leq s_1) + \sum_{l \in \mathcal{L}\backslash 0} P\left(s_l \leq \gamma_{m,k} \leq \min\left(\frac{\lambda\eta_l}{[w_m r_l - g_{m,k}^d]^+}, s_{l+1}\right)\right)$$

$$= u(g_{m,k}^d)F_{\gamma_{m,k}}(s_1) + \sum_{l \in \mathcal{L}\backslash 0} \left[F_{\gamma_{m,k}}\left(\min\left(\frac{\lambda\eta_l}{[w_m r_l - g_{m,k}^d]^+}, s_{l+1}\right)\right) - F_{\gamma_{m,k}}(s_l)\right]^+$$

Finally, the pdf (3.40) is the derivative of (D.2) with respect to $g_{m,k}^d$.

E

Derivation of (4.30)

Equating (4.29) to $\overline{\text{BER}}$ and after some algebraic manipulation, we have

$$\frac{\kappa_{m,k}}{\tilde{b}_{m,k}^{(l)} p_{m,k} + 1} e^{\left(\frac{\kappa_{m,k}}{\tilde{b}_{m,k}^{(l)} p_{m,k} + 1}\right)} = \frac{\kappa_{m,k}}{\tilde{a}_{m,k}^{(l)}} \overline{\text{BER}} \qquad (\text{E.1})$$

Observe that this is in the form of the Lambert-W function $W(x)$ [68], which is the solution to $W(x)e^{(W(x))} = x$. Thus, we can write

$$W\left(\frac{\kappa_{m,k}}{\tilde{a}_{m,k}^{(l)}} \overline{\text{BER}}\right) = \frac{\kappa_{m,k}}{\tilde{b}_{m,k}^{(l)} p_{m,k} + 1} \qquad (\text{E.2})$$

which when solved for $p_{m,k}$ gives us (4.30).

References

1. P. Rysavy. (2006) Mobile Broadband: EDGE, HSPA, and LTE. [Online]. Available: http://3gamericas.org/pdfs/white_papers/2006_Rysavy_Data_Paper_FINAL_09.15.06.pdf
2. H. Yaghoobi, "Scalable OFDMA Physical Layer in IEEE 802.16 WirelessMAN," *Intel Technology Journal*, vol. 8, no. 3, pp. 201–212, Aug 2004.
3. (2007) OFDM. [Online]. Available: http://en.wikipedia.org/wiki/OFDM
4. IEEE. (2007) Nikola Tesla, 1856-1943. [Online]. Available: http://www.ieee.org/web/aboutus/history_center/biography/tesla.html
5. telecomseurope.net. (2007) Underserved areas to drive Asia's telecom growth. [Online]. Available: http://www.telecomseurope.net/article.php?id_article=3754
6. S. Giodano and E. Biagioni, "Topics in ad hoc and sensor networks," *IEEE Commun. Mag.*, vol. 44, no. 11, p. 114, Nov 2006.
7. 3GAmericas. (2007) World Cellular Technology Forecast 2006-2011. [Online]. Available: http://www.3gamericas.org/English/Statistics/17.cfm
8. (2007) Total Midyear Population for the World: 1950-2050. [Online]. Available: http://www.census.gov/ipc/www/worldpop.html
9. (2007) WORLD INTERNET USAGE AND POPULATION STATISTICS. [Online]. Available: http://www.Internetworldstats.com/stats.htm
10. T. S. Rappaport, *Wireless communications : Principles and Practice*. Prentice Hall, 2002.
11. G. L. Stüber, *Principles of Mobile Communication*, 2nd ed. Kluwer Academic, 2001.
12. H. G. Myung, J. Lim, and D. J. Goodman, "Single carrier FDMA for uplink wireless transmission," *IEEE Vehicular Technology Magazine*, vol. 1, no. 3, pp. 30–38, 2006.
13. I. Wong, K. Oteri, and W. McCoy, "Optimal Resource Allocation in Uplink SC-FDMA Systems," *IEEE Trans. Wireless Commun.*, submitted.
14. C. Eklund, R. B. Marks, K. L. Stanwood, and S. Wang, "IEEE standard 802.16: a technical overview of the WirelessMAN air interface for broadband wireless access," *IEEE Commun. Mag.*, vol. 40, no. 6, pp. 98–107, 2002.
15. *IEEE Standard for Local and Metropolitan area networks Part 16: Air Interface for Fixed Broadband Wireless Access Systems*, IEEE Std. 802.16-2001, 2002.

16. *IEEE Standard for Local and Metropolitan Area Networks Part 16: Air Interface for Fixed Broadband Wireless Access Systems*, IEEE Std. 802.16-2004, 2004.
17. *Air Interface for Fixed and Mobile Broadband Wireless Access Systems*, IEEE Std. 802.16e-2005, Feb. 2006.
18. *Wireless LAN Medium Access Control (MAC) and Physical Layer (PHY) Specification*, IEEE Std. 802.11, 1997.
19. *3rd Generation Partnership Project, Technical Specification Group Radio Access Network; Physical layer aspects for evolved Universal Terrestrial Radio Access (UTRA)*, 3GPP Std. TR 25.814 v. 7.0.0, 2006.
20. T. Starr, J. Cioffi, and P. Silverman, *Understanding Digital Subscriber Line Technology*. Prentice Hall, 1999.
21. S. Baig and N. D. Gohar, "A discrete multitone transceiver at the heart of the PHY layer of an in-home power line communication local-area network," *IEEE Commun. Mag.*, vol. 41, no. 4, pp. 48–53, 2003.
22. J. Hayes, "Adaptive feedback communications," *IEEE Trans. Commun.*, vol. 16, pp. 29–34, Feb. 1968.
23. A. Goldsmith and S.-G. Chua, "Variable-rate variable-power MQAM for fading channels," *IEEE Trans. Commun.*, vol. 45, no. 10, pp. 1218–1230, Oct. 1997.
24. S. T. Chung and A. Goldsmith, "Degrees of freedom in adaptive modulation: a unified view," *IEEE Trans. Commun.*, vol. 49, no. 9, pp. 1561–1571, Sept. 2001.
25. R. Knopp and P. Humblet, "Information capacity and power control in single-cell multiusercommunications," in *IEEE International Conference on Communications*, vol. 1, 1995.
26. P. Viswanath, D. Tse, and R. Laroia, "Opportunistic beamforming using dumb antennas," *IEEE Trans. Inform. Theory*, vol. 48, no. 6, pp. 1277–1294, 2002.
27. T. Cover, "Broadcast channels," *IEEE Trans. Inform. Theory*, vol. 18, no. 1, pp. 2–14, Jan. 1972.
28. L. Li and A. Goldsmith, "Capacity and optimal resource allocation for fading broadcast channels Part I.-Ergodic capacity," *IEEE Trans. Inform. Theory*, vol. 47, no. 3, pp. 1083–1102, Mar. 2001.
29. ——, "Capacity and optimal resource allocation for fading broadcast channels .II. Outage capacity," *IEEE Trans. Inform. Theory*, vol. 47, no. 3, pp. 1103–1127, Mar. 2001.
30. A. Goldsmith and M. Effros, "The capacity region of broadcast channels with intersymbol interference and colored Gaussian noise," *IEEE Trans. Inform. Theory*, vol. 47, no. 1, pp. 219–240, Jan. 2001.
31. N. Jindal and A. Goldsmith, "Capacity and optimal power allocation for fading broadcast channels with minimum rates," *IEEE Trans. Inform. Theory*, vol. 49, no. 11, pp. 2895–2909, Nov. 2003.
32. C. Y. Wong, R. Cheng, K. Lataief, and R. Murch, "Multiuser OFDM with adaptive subcarrier, bit, and power allocation," *IEEE J. Select. Areas Commun.*, vol. 17, no. 10, pp. 1747–1758, Oct. 1999.
33. B. S. Krongold, K. Ramchandran, and D. L. Jones, "Computationally efficient optimal power allocation algorithms for multicarrier communication systems," *IEEE Trans. Commun.*, vol. 48, no. 1, pp. 23–27, 2000.
34. D. Kivanc, G. Li, and H. Liu, "Computationally efficient bandwidth allocation and power control for OFDMA," *IEEE Trans. Wireless Commun.*, vol. 2, no. 6, pp. 1150–1158, Nov. 2003.

35. I. Kim, H. L. Lee, B. Kim, and Y. Lee, "On the use of linear programming for dynamic subchannel and bit allocation in multiuser OFDM," in *Proc. IEEE Global Telecommunications Conference*, vol. 6, Nov. 2001, pp. 3648–3652.

36. M. Ergen, S. Coleri, and P. Varaiya, "QoS aware adaptive resource allocation techniques for fair scheduling in OFDMA based broadband wireless access systems," *IEEE Trans. Broadcast.*, vol. 49, no. 4, pp. 362–370, Dec. 2003.

37. J. Jang and K. B. Lee, "Transmit Power Adaptation for Multiuser OFDM Systems," *IEEE J. Select. Areas Commun.*, vol. 21, pp. 171–178, Feb. 2003.

38. J. Jang, K. B. Lee, and Y.-H. Lee, "Frequency-time domain transmit power adaptation for OFDM systems in multiuser environment," *IEE Electronics Letters*, vol. 38, no. 25, pp. 1754–1756, 2002.

39. H. Yin and H. Liu, "An Efficient Multiuser Loading Algorithm for OFDM-based Broadband Wireless Systems," in *Proc. IEEE Global Telecommunications Conference*, vol. 1, Dec. 2000, pp. 103–107.

40. Y. Zhang and K. Letaief, "Multiuser subcarrier and bit allocation along with adaptive cell selection for OFDM transmission," in *Proc. IEEE International Conference on Communications*, vol. 2, Apr. 2002, pp. 861–865.

41. W. Rhee and J. M. Cioffi, "Increase in Capacity of Multiuser OFDM System Using Dynamic Subchannel Allocation," in *Proc. IEEE Vehicular Technology Conference*, Tokyo, Japan, May 2000, pp. 1085–1089.

42. L. Hoo, B. Halder, J. Tellado, and J. Cioffi, "Multiuser transmit optimization for multicarrier broadcast channels: asymptotic FDMA capacity region and algorithms," *IEEE Trans. Commun.*, vol. 52, no. 6, pp. 922–930, June 2004.

43. K. Seong, M. Mohseni, and J. Cioffi, "Optimal resource allocation for OFDMA downlink systems," in *Proc. IEEE International Symposium on Information Theory*, Seattle, WA, July 2006, pp. 1394–1398.

44. Y. Yu, X. Wang, and G. B. Giannakis, "Channel-adaptive congestion control and OFDMA scheduling for hybrid wireline-wireless networks," *IEEE Trans. Wireless Commun.*, submitted for publication.

45. Z. Shen, J. Andrews, and B. Evans, "Adaptive resource allocation in multiuser OFDM systems with proportional rate constraints," *IEEE Trans. Wireless Commun.*, vol. 4, no. 6, pp. 2726–2737, Nov. 2005.

46. I. C. Wong, Z. Shen, B. Evans, and J. Andrews, "A low complexity algorithm for proportional resource allocation in OFDMA systems," in *Proc. IEEE Workshop on Signal Processing Systems*, Oct. 2004, pp. 1–6.

47. T. C. H. Alen, A. Madhukumar, and F. Chin, "Capacity enhancement of a multi-user OFDM system using dynamic frequency allocation," *IEEE Trans. Broadcast.*, vol. 49, no. 4, pp. 344–353, Dec. 2003.

48. Z. Han, Z. Ji, and K. J. R. Liu, "Fair Multiuser Channel Allocation for OFDMA Networks Using Nash Bargaining and Coalitions," *IEEE Trans. Commun.*, vol. 53, no. 8, pp. 1366–1376, Aug. 2005.

49. G. Li and H. Liu, "Dynamic resource allocation with finite buffer constraint in broadband OFDMA networks," in *Proc. IEEE Wireless Communications and Networking Conf.*, vol. 2, Mar. 2003, pp. 1037–1042.

50. S. Kittipiyakul and T. Javidi, "Subcarrier allocation in OFDMA systems: beyond water-filling," in *Proc. IEEE Asilomar Conference on Signals, Systems and Computers*, vol. 1, Nov. 2004, pp. 334–338.

51. H. Seo and B. G. Lee, "A proportional-fair power allocation scheme for fair and efficient multiuser OFDM systems," in *Proc. IEEE Global Telecommunications Conference*, vol. 6, Dec. 2004, pp. 3737–3741.

52. Z. Zhang, Y. He, and E. K. P. Chong, "Opportunistic downlink scheduling for multiuser OFDM systems," in *Proc. IEEE Wireless Communications and Networking Conference*, New Orleans, LA, USA, Mar. 2005, pp. 1206–1212.

53. X. Liu, E. Chong, and N. Shroff, "Opportunistic transmission scheduling with resource-sharing constraints in wireless networks," *IEEE J. Select. Areas Commun.*, vol. 19, no. 10, pp. 2053–2064, Oct. 2001.

54. G. Song and Y. Li, "Cross-Layer Optimization for OFDM Wireless Networks Part I: Theoretical Framework," *IEEE Trans. Wireless Commun.*, vol. 4, no. 2, pp. 614–624, Mar. 2005.

55. ——, "Cross-Layer Optimization for OFDM Wireless Networks Part II: Algorithm Development," *IEEE Trans. Wireless Commun.*, vol. 4, no. 2, pp. 625–634, Mar. 2005.

56. G. Song, Y. Li, and J. L. J. Cimini, "Joint channel- and queue-aware scheduling for multiuser diversity in wireless multicarrier networks," *IEEE Trans. Inform. Theory*, submitted for publication.

57. D. P. Bertsekas, *Nonlinear Programming*, 2nd ed.   Athena Scientific, 1999.

58. W. Yu and R. Lui, "Dual methods for nonconvex spectrum optimization of multicarrier systems," *IEEE Trans. Commun.*, vol. 54, no. 7, pp. 1310–1322, July 2006.

59. X. Wang, G. Giannakis, and A. Marques, "A unified approach to QoS-Guaranteed Scheduling for Channel-Adaptive Wireless Networks," *Proc. IEEE*, submitted.

60. I. C. Wong and B. L. Evans, "Optimal OFDMA resource allocation with linear complexity to maximize ergodic weighted sum capacity," in *Proc. IEEE Int. Conf. on Acoustics, Speech, and Signal Processing*, Honolulu, HI, April 2007.

61. ——, "Optimal Downlink OFDMA Subcarrier, Rate, and Power Allocation with Linear Complexity to Maximize Ergodic Weighted-Sum Rates," in *Proc. IEEE Int. Global Communications Conf.*, Washington, DC USA, submitted.

62. ——, "Optimal Downlink OFDMA Resource Allocation with Linear Complexity to Maximize Ergodic Rates," *IEEE Trans. Wireless Commun.*, 2006 *to appear*.

63. Y. Ermoliev and R. Wets, *Numerical Techniques for Stochastic Optimization*. Springer-Verlag, 1988.

64. D. Luenberger, *Optimization by Vector Space Methods*.   John Wiley and Sons, 1969.

65. D. R. Smith, *Variational Methods in Optimization*.   Prentice-Hall, 1974.

66. W. H. Press, *Numerical Recipes in C*.   Cambridge University Press Cambridge, 1992.

67. H. A. David and H. N. Nagaraja, *Order Statistics*, 3rd ed.   John Wiley, 2003.

68. R. Corless, G. Gonnet, D. Hare, D. Jeffrey, and D. Knuth, "On the Lambert-W function," *Advances in Computational Mathematics*, vol. 5, no. 1, pp. 329–359, 1996.

69. A. Papoulis and S. U. Pillai, *Probability, Random Variables, and Stochastic Processes*.   McGraw-Hill, 2002.

70. D. P. Bertsekas, *Constrained Optimization and Lagrange Multiplier Methods*. Academic Press, 1982.

71. *Selection procedures for the choice of radio transmission technologies for the UMTS*, ETSI Std. TR 101 112 v. 3.2.0, 1998.

72. I. C. Wong and B. L. Evans, "Optimal OFDMA subcarrier, rate, and power allocation for ergodic rates maximization with imperfect channel knowledge,"

in *Proc. IEEE Int. Conf. on Acoust., Sp., and Sig. Proc.*, Honolulu, HI, April 2007.

73. ——, "OFDMA Resource Allocation for Ergodic Capacity Maximization with Imperfect Channel Knowledge'," in *Proc. IEEE Int. Global Communications Conf.*, Washington, DC USA, submitted.

74. ——, "Optimal Downlink Resource Allocation in OFDMA Systems with Imperfect Channel Knowledge," *IEEE Trans. Commun.*, 2006 *submitted for publication*.

75. D. Goeckel, "Adaptive coding for time-varying channels using outdated fading estimates," *IEEE Trans. Commun.*, vol. 47, no. 6, pp. 844–855, 1999.

76. S. Falahati, A. Svensson, T. Ekman, and M. Sternad, "Adaptive modulation systems for predicted wireless channels," *IEEE Trans. Commun.*, vol. 52, no. 2, pp. 307–316, Feb. 2004.

77. M. R. Souryal and R. L. Pickholtz, "Adaptive Modulation with Imperfect Channel Information in OFDM," in *Proc. IEEE Int. Conf. Comm.*, June 2001, pp. 1861–1865.

78. S. Ye, R. S. Blum, and J. L. J. Cimini, "Adaptive modulation for variable-rate OFDM systems with imperfect channel information," in *Proc. IEEE Vehicular Technology Conference*, vol. 2, Spring 2002, pp. 767–771.

79. I. C. Wong, A. Forenza, R. Heath, and B. Evans, "Long range channel prediction for adaptive OFDM systems," in *Proc. IEEE Asilomar Conf. on Sig., Sys., and Comp.*, vol. 1, Nov. 2004, pp. 732–736.

80. I. C. Wong and B. L. Evans, "Joint Channel Estimation and Prediction for OFDM Systems," in *Proc. IEEE Global Telecommunications Conference*, St. Louis, MO, Dec. 2005, pp. 2255–2259.

81. ——, "Sinusoidal Modeling and Adaptive Channel Prediction for OFDM Systems," *IEEE Trans. Signal Processing*, submitted.

82. I. C. Wong and B. Evans, "Performance Bounds in OFDM Channel Prediction," in *Proc. IEEE Conference on Signals, Systems and Computers*, Pacific Grove, CA, USA, Oct. 28 - Nov. 1, 2005, pp. 1461–1465.

83. I. C. Wong and B. L. Evans, "Low-Complexity Adaptive High-Resolution Channel Prediction for OFDM Systems," in *Proc. IEEE Global Telecom. Conf.*, San Francisco, CA, Nov 2006.

84. Y. Yao and G. Giannakis, "Rate-Maximizing Power Allocation in OFDM Based on Partial Channel Knowledge," *IEEE Trans. Wireless Commun.*, vol. 4, no. 3, pp. 1073–1083, May 2005.

85. I. C. Wong and B. L. Evans, "Exploiting Spatio-Temporal Correlations in MIMO Wireless Channel Prediction," in *Proc. IEEE Global Telecommunications Conf.*, San Francisco, CA, Nov Wong06b, p. accepted for publication.

86. S. Zhou and G. Giannakis, "How accurate channel prediction needs to be for transmit-beamforming with adaptive modulation over Rayleigh MIMO channels?" *IEEE Trans. Wireless Commun.*, vol. 3, no. 4, pp. 1285–1294, July 2004.

87. P. Xia, S. Zhou, and G. Giannakis, "Adaptive MIMO-OFDM based on partial channel state information," *IEEE Trans. Signal Processing*, vol. 52, no. 1, pp. 202–213, Jan. 2004.

88. J. G. Proakis, *Digital communications*, 4th ed.   McGraw-Hill, 2001.

89. I. S. Gradshteyn, I. M. Ryzhik, and A. Jeffrey, *Table of Integrals, Series, and Products*, 6th ed.   Academic Press, 2000.

90. M. Abramowitz and I. A. Stegun, *Handbook of Mathematical Functions with Formulas, Graphs, and Mathematical Tables*, 10th ed.   U.S. Govt. Print. Off., 1972.

91. Z. Shen, J. G. Andrews, and B. L. Evans, "Optimal Power Allocation in Multiuser OFDM Systems," in *Proc. IEEE Global Telecommunications Conference*, vol. 1, Dec. 2003, pp. 337–341.

92. W. Yu, R. Lui, and R. Cendrillon, "Dual optimization methods for multiuser orthogonal frequency division multiplex systems," in *Proc. IEEE Global Telecommunications Conference*, vol. 1, Dec. 2004, pp. 225–229.

93. S. Ross, *Simulation*, 3rd ed.   Academic Press, 2002.

94. H. J. Kushner and G. Yin, *Stochastic Approximation and Recursive Algorithms and Applications*, 2nd ed.   Springer, 2003.

95. J. R. Birge and F. Louveaux, *Introduction to Stochastic Programming.* Springer, 1997.

96. H. Robbins and S. Monro, "A stochastic approximation method," *Ann. of Math. Statist.*, vol. 22, pp. 400–407, 1951.

97. B. T. Polyak and A. B. Juditsky, "Acceleration of stochastic approximation by averaging," *SIAM J. Control Optim.*, vol. 30, no. 4, pp. 838–855, 1992.

98. A. Ruszczynski and W. Syski, "Stochastic Approximation Method with Gradient Averaging for Unconstrained Problems," *IEEE Trans. Automat. Contr.*, vol. 28, no. 12, pp. 1097–1105, Dec 1983.

99. A. Paulraj, R. Nabar, and D. Gore, *Introduction to space-time wireless communications.*   Cambridge University Press, 2003.

Index